Response Threshold Based Task Allocation in Multi-Agent Systems Performing Concurrent Benefit Tasks with Limited Information

Anshul Kanakia

Dedication

I would like to dedicate this thesis to my late grandfather Mr. Manilal Kanakia (1923–2014). A self made man with humble beginnings from a small village in Gujarat, India, he ended his career as an emeritus school principal, academic, author, and editor of several Geography text books. He taught me the true meaning of patience and the invaluable skill of honest ingenuity — maximizing benefit from the fewest resources possible.

As is perhaps the norm this year of 2015, I would also like to dedicate my Thesis to the logically infallible Mr. Leonard Nimoy, a.k.a. "Mr. Spock" (1931–2015). His life's work will forever inspire humanity towards a utopia where we may all one day live long and prosper.

Acknowledgments

I am eternally grateful to my colleagues and friends at the University of Colorado, Correll Robotics Lab. Much of my inspiration for this work has come from entertaining and enlightening conversations with my lab buddies, plus countless hours of Halo and Tetris in their company! Of course, none of this work would have been possible without the incredible mentoring of my adviser Dr. Nikolaus Correll as well as my thesis committee. Finally, I would like to confess my eternal gratitude and love for my beautiful wife, Mariah. Even though grad school kept us apart for many years your loving thoughts and enduring patience gave me all the motivation I could have asked for!

Thesis Committee

- Dr. Nikolaus Correll. CU Boulder, Computer Science **(Chair)**

- Dr. Sriram Sankaranarayanan. CU Boulder, Computer Science

- Dr. Ani Hsieh. Drexel University, Computer Science

- Dr. Behrouz Touri. CU Boulder, Electrical and Computer Engineering

- Dr. Gabe Sibley. CU Boulder, Computer Science

Abstract

One of the most elusive but important goals of swarm robotics is to reproduce the emergent collaborative behavior observed in natural swarming systems through the use of simple decision rules. Examples of collaborative processes in insect colonies such as foraging, scouting (finding shortest paths) for food, and colony defense involve some form of task allocation among individual agents. The robustness of task completion even after major environmental changes is also observed in natural swarm systems. Ants and bees are often unphased by the fact that the magnitude of a task — such as carrying a heavy piece of food — is unknown to every individual and manage to complete the task elegantly even without such critical knowledge. This robustness property is of paramount importance when recreating natural behavior in artificial systems and I believe the use of decentralized agent based task allocation rules is closely related to this property. I therefore present a novel response threshold based strategy for task allocation in multi-agent systems in this dissertation. I prove, using a well known result from the theory of global games, that under the constraints of imperfect knowledge of the environment and imperfect communication response threshold based task allocation leads to an equilibrium inducing strategy for the swarm system. The importance of this result is to provide a formal mathematical basis for the phenomenological justification currently provided in the field of swarm robotics to mimic biological systems. This result therefore provides both, a hypothesis about the inner workings of a wide range of existing approaches with limited communication between agents in artificial swarm systems and also a formal explanation for threshold based task allocation in social insects. These game theory results lead to a novel continuous response threshold algorithm for multi-agent task allocation that generalizes fixed-group task allocation (stick-pulling experiment) and stochastic team size task allocation. This allows variable team sizes to form at task sites within tolerance limits thereby providing a trade-off between exploration and exploitation. The claims made by theoretical proofs for response threshold based task allocation are backed up by physical experiments using the *Droplet* swarm robot platform. Further simulation experiments provide a basis of comparison between optimal centralized approaches and hybrid approaches for task allocation where each robot decides whether to participate in a task based on its own noisy sensory input and imperfect knowledge from the system controller. I show that in many real world situations it is often impractical to rely on the assumption of perfect system information for controlling a swarm and that centralized task allocation becomes comparable to a response threshold based policy under the influence of noise.

Acronyms

MAS   Multi-Agent System
TA   Task Allocation
MATA   Multi-Agent Task Allocation
RT   Response Threshold
CRT   Continuous Response-Threshold
DRT   Discrete Response-Threshold
ODE   Ordinary Differential Equation
FSM   Finite State Machine
PFSM   Probabilistic Finite State Machine

Symbols Used

Symbol	Description
$\mathcal{P}$	Set of agents.
$\mathcal{T}$	Set of targets or tasks.
n_i	Single agent i.
t_i	Single target i.
τ	Target magnitude. τ_i is the magnitude of task i. This may be a constant or a function.
x_i	Target magnitude for target i with Gaussian noise added.
k_i	Target threshold. This is the minimum number of agents required to successfully complete a task. Closely related to task magnitude τ_i. A function can map from one to the other.
u_i	Agent level utility. This may be a constant or a function.
$\mathcal{A}(t_j)$ or α_j	Set of agents assigned to target j.
$W(t_j)$	System welfare for completing a target j.
$\mathcal{N}(\mu, \sigma)$	Normal or Gaussian distribution with mean μ and std. deviation σ.

Contents

1 Introduction   **13**
   1.1   Related Work   15
   1.2   Main Contribution and Overview   17

2 Background on Swarm System Modeling Techniques   21
   2.1   Terms   22
       2.1.1   Agent or Robot   22
       2.1.2   Stigmergy   22
       2.1.3   Swarm System   23
       2.1.4   Microscopic Level   23
       2.1.5   Macroscopic Level   24
   2.2   Designing the Controller Construct   24
   2.3   Mathematical Description of the
        System   27
   2.4   Microscopic Simulation of the
        System   29
   2.5   Verification of System Properties Using Real Ex-
        periments and
        Physics-Based Simulation   30
   2.6   Summary   32

**3 Designing an Optimal Control Model for Multi-
  Agent Systems   33**
   3.1   Multi-Agent Task Allocation Model   34
   3.2   Defining Optimality for
        Multi-Agent Task Allocation   37

4 Existence of an Equilibrium Strategy for Communication Free Multi-Robot Task Assignment   **39**
   4.1 Global Games: A Brief Overview 40
   4.2 Task Allocation as Global Games 41
   4.3 Communication Free Threshold
      Based Task Allocation Strategy 45
   4.4 From Discrete Thresholds to
      Sigmoidal Response Functions 48
   4.5 Discussion and Summary 50

5 Response Threshold Model for Multi-Agent Task Allocation   **53**
   5.1 Macroscopic analysis 56
   5.2 Microscopic Model 57
   5.3 Experiments and Results 59
   5.4 Discussion 63
   5.5 Summary 66

6 Comparing Centralized vs. Hybrid Approaches to Task Allocation   **67**
   6.1 Centralized Optimal Allocations 68
   6.2 Distributed Response Threshold
      based Task Allocation 70
   6.3 Experiments 71
   6.4 Results 74
   6.5 Discussion 76

7 Conclusion   **79**

Chapter 1

Introduction

Many multi-agent social, biological and physical systems can be categorized as swarm intelligence [1]. From modeling human social interaction and population dynamics to insect colonies and great animal herd migrations, from cellular automata to distributed network systems and even the interconnected computing devices that form the internet can be viewed as swarm intelligence. Swarm robotics [2] is a branch of swarm intelligence applied to physical multi-agent systems (MAS) to leverage the advantage of producing emergent, complex behavior from individually simplistic agents and rules. This has led to novel approaches in the design and analysis of MAS and the algorithms associated with them [3]. Swarm robotics has tackled a vast array of MAS problems in the past two decades. It's corpus ranges from self-organization, self-assembly, pattern formation, and aggregation to foraging, coordinated movement (such as flocking and schooling), collective transport, and group surveillance. Readers are directed to [4] and the references therein for further information on any of these topics.

Swarm systems have many benefits over traditional, centralized robot systems. The robots used in swarm applications are generally many orders of magnitude smaller (cm vs. m, grams vs. kg) and simpler in design (< 10 vs. 100s of actuators) than conventional robots, while being much greater in number (10^2 to 10^{23}). Also, most swarm systems are homogeneous—robots with identical software/hardware are used to complete the assigned task. This makes swarm systems easily scalable while simultaneously

Figure 1.1: The *Droplet* swarm robots running a fire containment experiment inspired by a real forest firefighting scenario.

keeping manufacturing and maintenance costs of the hardware low. Though, perhaps their greatest advantage is system stability and robustness to error. Most swarm systems consist of small, relatively simple robots that are only capable of limited and noisy sensing, communication and actuation. This means that while no single robot alone is capable of performing the task assigned, the system as a whole is resilient to individual unit errors and is capable of completing the task [5].

Performing collaborative tasks is a vast sub-field of study in swarm robotics and considerable work has been done to understand and model such scenarios, particularly by [6–10] using the well known stick-pulling experiments (The reader is directed to the Related Work section for more information on the stick-pulling experiment if they are unfamiliar with it). Collaborative tasks using MAS extend to a variety of potential real-world applications such as object transport [11], oil-spill containment [1], firefighting [12], collective inspection [13], pattern recognition [14], and cooperative surveillance. In such cases it is impractical and often impossible to know beforehand, exactly how many agents are required for successful completion. More importantly, the benefit of larger teams of agents for such collaborative tasks increases non-linearly with team size, e.g. Where 1-5 robots may be incapable of lifting a heavy object, 6-7 will be able to lift and move it successfully but the usefulness more than 7 robots begins to diminish rapidly thereafter. I define such tasks with non-linearity in system utility with varying team sizes as *concurrent benefit* tasks. The original work presented in this dissertation applies to concurrent benefit tasks and particularly, the dynamics governing variable team size task assignment (TA) for concurrent benefit tasks under noisy communication and sensing constraints. Empirical evidence from biological systems suggests that there exists a generalized solution for collaborative TA in MAS. This solution involves the use of response threshold (RT) functions which leads to a system-level equilibrium strategy that can be used for robust decentralized control of a MAS.

1.1 Related Work

TA is a canonical problem in MAS research [15]. While capable robots might be able to approximate optimal TA, e.g., using market-based approaches [16, 17] or using leader-follower coalition algorithms [18], probabilistic algorithms are of particular interest for swarm robotics with individually simple controllers [19]. Recruitment of an exact number of robots to a particular task has been extensively studied using the stick-pulling experiment [20, 21]. The problem of distributing a swarm of robots across a discrete number of sites/tasks with a specific desired distribution has been studied in [22, 23]. Mather [24] instead presents a stochastic approach that is a hybrid between the work in [22] and [21], allowing allocation to tasks requiring a varying number of robots. The RT based TA algorithm presented in this dissertation extends upon the first group of work. I show how the RT TA algorithm reduces to the ones described in [20, 21] when using appropriate parameters.

Using RT functions to model social behavior in insects such as ant colonies [25, 26] and bee hives [27–29] has been proposed, analyzed and verified by biologists since the 1980's [30]. In the past two decades swarm roboticists have begun to engineer MASs using these models. Jones and Matarić [31] describe an adaptive method of TA for a large-scale minimalist robot system where agents independently switch between picking up different colored pucks to maintain a consistent rate of foraging for each type of colored puck. While the authors do not directly reference threshold functions, their switching algorithm simply assigns probabilities of picking up a certain colored puck versus another by accounting for the number of colored pucks observed by a robot around it, which is a form of probabilistic threshold policy. Such dynamic probabilistic threshold policies are also studied in [32]. I proposed a modification of such strategies to instead use the logistic sigmoid function [12]. Using a logistic function better exposes mean and variance parameters of the resulting team sizes for CRT functions. These two parameters, along with the number of workers available and the manner in which they acquire information, are the building blocks governing TA behavior of an agent in the collective swarm (see section titled, "Division of Labor as a Self-Organizing Process" in [28]).

All of the aforementioned work uses continuous response threshold (CRT) functions. In contrast to this, discrete response thresholds (DRTs, such as step-functions) for TA have been studied by a number of research groups utilizing the stick-pulling experiment [6,20,21,33]. Here, an enclosed arena is set up with a number of sticks or *task sites* that must be pulled out of the ground by a team of two or more robots working collaboratively. Robots arrive at a stick and wait a certain amount of time for a partner to arrive and assist in the stick pulling process. The ratio between the number of sticks to the number of robots in the arena creates interesting dynamics where the stick pulling rate can by maximized by optimizing wait times if there are fewer robots than sticks. Here, a task can only be solved with an exact number of robots $= \Upsilon$. The problem of distributing a swarm of robots across multiple sites with a specific desired distribution has been studied in [22,23] and is extended by Mather [24] allowing assignment to tasks requiring a varying number of robots. The benefits of using a TA algorithm versus just allowing agents to attempt a collaborative task, such as aggregation, without a RT is analyzed in [8]. The authors show that threshold based TA results in increasing aggregation of seeds while no TA results in stagnation of seed collection after a little while.

The development of MATA draws a very clear picture of its evolution from a behavioral model for insect colonies, developed by ethologists, to an inspired algorithmic model for adaptive MASs

[34]. While biologists have provided ample empirical evidence to the success of the RT model in predicting and matching observed swarm behavior for TA, there has been no formal argument as to why natural systems gravitate towards this approach compared to other TA strategies such as leader-follower algorithms [18] or market-based approaches [16,17]; see [35] for a comparative study. My aim is to show for the first time that agent-level RT strategies drive a swarm system to some notion of equilibrium and consequently, system-level control which makes them an obvious choice for modeling natural systems and engineering artificial ones.

1.2   Main Contribution and Overview

It is clear that MAS TA is an important problem to solve if we want robots to one day collaboratively complete complex tasks in a distributed manner without human intervention even under noisy sensory conditions. With this goal in mind I present my RT based TA strategies for MAS.

Chapter 2 provides an outline of existing swarm modeling techniques, particularly multi-level abstraction using probabilistic finite state machines (PFSMs) and rate equations. Terms commonly used throughout this dissertation are also defined here. Readers familiar with swarm modeling techniques may skip this tutorial chapter and move straight to Chapter 3.

The MATA model used throughout this dissertation is presented in Chapter 3. The model presented in this chapter is inspired from the field of utility theory and is general enough to encompass a wide array of more task specific multi-agent allocation models used in swarm robotics. It establishes the concept of task centric utility and lays the foundation for a discussion on TA *optimality* in multi-agent systems. Later, this concept of optimality resurfaces in Chapter 6 where I experimentally compare and analyze centralized vs. distributed TA controllers.

In Chapter 4, the focus shifts to finding distributed strategies for MATA. I choose to focus on a specific distributed TA strategy called a response threshold strategy. The concept of global games from Game Theory lends an important result proving the existence of a Bayesian Nash equilibrium (BNE) when using threshold based strategies (as opposed to other distributed methods for TA) in systems where agents have imperfect information about the environment. The proof of the above statement forms the basis for Chapter 4. This chapter also provides another theorem generalizing DRT based models to CRT models for TA. Much of the work presented in this chapter is through collaborative efforts with Dr. Behrouz Touri. Specifically, the proofs presented in this chapter present a substantial contribution from Dr. Touri towards my work and end goals. Due to a heavy reliance on game theory formalism for proofs of the two theorems, Section 4.1 of Chapter 4 provides a brief introduction to the theory of global games and can be skipped by readers familiar with the material.

Extending upon the global game theorems, Chapter 5 experimentally shows that the DRT strategy is a special case of CRTs when the slope of the response threshold functions approach infinity, and thus the step-function behavior can be accurately reproduced by the latter. In practice, varying the slope of the RT function allows one to balance exploration and exploitation in the system [26]. In this chapter I present my novel response threshold model of MATA that unifies existing deterministic and probabilistic approaches into a single tunable model. The CRT model involves agents using a logistic function with two tunable parameters to control the mean and variance of desired team size formation. I analyze and verify this model using real robot experiments as well as simulation results. It is worth noting that biological systems do not necessarily implement sigmoid functions but that they might naturally emerge from a combination of a DRT and noise in the perception system, which I show analytically in Chapter 4.

Finally, the response threshold model for TA from the previous chapter is compared to an *optimal* centralized strategy in Chapter 6. This final chapter brings together all my contributions, from the formal definition for optimal mult-agent TA in Chapter 3 to the logistic function based CRT controller from Chapter 5. It analyses what happens when noisy centralized controllers are used in tandem with noisy distributed control and verifies the notion that a hybrid approach to TA is desirable in most real-world scenarios where perfect information about the system cannot be discerned.

Chapter 1. Introduction

The conclusion lays out areas for possible future work in MATA and how the controllers described in this dissertation can possibly be extended to robustly handle real-world scenarios where large teams of robots could collaboratively perform group tasks without complex control algorithms.

Chapter 1. Introduction

Chapter 2

Background on Swarm System Modeling Techniques

This chapter is designed to give the reader an introductory lesson on swarm system modeling approaches widely used by the swarm robotics community. Terms used commonly throughout this paper are also defined in the next section. Modeling robot swarms often involves the use of PFSMs to mathematically describe individual agent behavior as well as solving systems of coupled ODEs, often termed *rate equations*, to divine macroscopic level properties of the system. These are common tools of the trade for modeling many different kinds of dynamical systems—not just MAS—and as such, if the reader is familiar with these approaches they are encouraged to skip this chapter and move on to Chapter 3.

The first step in the robot controller design methodology is to describe the swarming task being studied. The general strategy used by each individual agent in the swarm is defined and later translated into a viable agent-level, or microscopic model. A hypothesis for the observed, collective behavior of the swarm is supplied, which is later quantified into a system-level or macroscopic model.

Physical characteristics of the swarm system are generally described in the experiment setup as well. These may include environmental variables such as arena size, agents' properties such as speed, communication and sensing radii, the computation power of each individual in the swarm, etc. This is an important step in identifying important system parameters that affect the outcome of the experiment, versus the environmental and agent based values that can be abstracted away when designing models at different levels of abstraction.

2.1 Terms

2.1.1 Agent or Robot

Russel and Norwig [36] define an agent or robot as "Anything that can be viewed as perceiving its environment through sensors and acting upon that environment through effectors." While this definition is sufficient in most cases, it excludes a swarm system's propensity for communication. Therefore, an enhanced definition for a swarm robot could be:

> Any mechanical automation capable of sensing it's surroundings, processing sensory inputs via internal computation, actuating itself or other objects in the environment based on the inputs, and communicating information with other robots around it, either directly or via *stigmergy*.

2.1.2 Stigmergy

Stigmergy is a term used to describe indirect communication in robot swarms. The word was first coined by Pierre-Paul Grassé, in his 1959 paper [37], while studying insect behavior and has become a commonly used term also in the swarm robotics community.

While most robots are capable of communicating amongst themselves explicitly via infrared, Bluetooth^{TM}, and other wired or wireless means, most swarm algorithms try to keep such explicit communication to a minimum. This is done to maintain scalability of the system (e.g., prevent message flooding) and keep the underlying algorithm simple.

Indirect methods of communication are thus preferred when designing controllers for robot swarms, such as changing one's color or moving in a particular pattern or even just waiting at a specific position in the environment. The process of adding information to the swarm system by affecting or altering the environment rather than explicitly communicating with other agents is referred to as a stigmergic process [38]. It is used in many of the swarm algorithms discussed in this paper.

2.1.3 Swarm System

A myriad of definitions are available for systems and models that exhibit "swarming" phenomena—the term "swarming" itself being re-defined in numerous cases. An interesting discussion of the nomenclature of swarm systems is available in [1, 39].

Since our primary purpose in modeling robot swarms is to study and understand properties of the system as a whole, the term *swarm system* or *multi-agent system* (MAS) is not used just as a collective noun for a group of robots but instead describes the complex relationship between agents, the environment, and the tasks they are attempting to accomplish.

Multi-agent systems can be modeled at different levels of abstraction depending on the system properties we attempt to expose. When these models are used in tandem we gain the ability to both, verify and enhance the original swarm system via optimization of system parameters. We see the deployment of this strategy in the next section but first, we define the different abstraction levels used in robot swarm modeling.

2.1.4 Microscopic Level

The micro-level of a MAS model treats the individual agent as the fundamental unit of the model [40]. Though not a requirement for this form of agent-based modeling, we generally assume that the swarm is homogeneous, i.e. every agent is running the same robot controller within it and all hardware (sensors, actuators, processors and communication devices) between robots is identical. The microscopic level then helps describe direct agent-agent interactions as well as agent-environment interaction.

An example of micro-level modeling consists of writing down the dynamics equations (equations of motion) for an individual robot and solving them to study system-level behavior. As one can imagine, these dynamics equations can become very tedious and difficult to solve for more complex swarm systems due to the high number of agents, inelastic collisions between agents and obstacles, sensor and actuator noise, etc. Therefore a more common approach to micro-level modeling involves stochastic simulation of individual robot controllers in parallel, ignoring individual trajectories and positions. This micro-level modeling method has been derived from the popular *Gillespie* method [41, 42] used to model coupled chemical reactions. Another example of microscopic model—with a much lower level of abstraction—is a simulator that simulates the motion and possibly collisions among robots.

2.1.5   Macroscopic Level

While micro-level models deal with systems on an individual agent level, macro-level models consider the system as a whole,and are used to describe collective group behavior. Macro-models for robot swarms are often phenomenological in nature. The system's parameters are derived from observing and measuring real physical phenomena and extrapolating such properties as may be deemed useful for understanding said phenomena. Macro-level models are generally represented as a system of ordinary (for non-spatial models) or partial (for spatial models) differential equations and as such are good at describing the temporal and spatial evolution of the system. They are often also referred to as population dynamics models and/or rate equations in literature.

2.2   Designing the Controller Construct

Creating a logic construct—a flowchart, state-machine or algorithm—that describes the desired robot behavior for the given task is an important step in the MAS modeling process. When studying non-spatial models, the robot controller can be characterized by an FSM with a discrete number of states under urgent time-step driven semantics, as seen in Figure 2.1. The states in the FSM (s_1, s_2, s_3 & s_4) represent physical states that the robot can be in, such as *searching, waiting*, etc. and can be directly derived from the program code running on the robot. One can think of each state as being an *action* that the robot is currently performing based on stimulus from the environment and other robots. These stimuli can cause a robot to transition from one state to another and are represented as *conditionals* on the edges of the FSM, c_i. These conditionals are equivalent to the decision process blocks in the flowchart and can be derived from:

1. Sensor readings (or stigmergy) and explicit communication, e.g., seeing red light through an rgb sensor or seeing a certain number of robots around you,

2. internal timers, e.g., transition back to search after waiting for 3 seconds,

3. or a combination of both, e.g., transition back to search after waiting for 3 seconds *iff* you see no other robots in your vicinity, otherwise, reset your timer.

I can now extend this modeling framework of robot behavior as an FSM to construct a PFSM, where the conditionals in the FSM are no longer true/false values but instead are probabilities of transitioning from one state to another based on external stimulus or internal state.

As alluded to earlier, the case of a state transition based on an internal timer is especially interesting. Let c_1 in Figure 2.1a be the condition $t_{s_1} \geq 5$, i.e. time in state s_1 is greater than or equal to 5 time steps (or ticks). This conditional is true when the robot has remained in state s_1 for at least 5 ticks and consequently transitions to state s_2. Thus, the conditional c_1 says a robot may remain in state A for no more than 5 ticks. The equivalent transition probability for this condition would be $p_{s_1} = 1/5$. Therefore, at each time step of the PFSM simulation, there is a $1/5$ chance that the robot will transition from state A to state B. The expected number of ticks before a transition happens is then

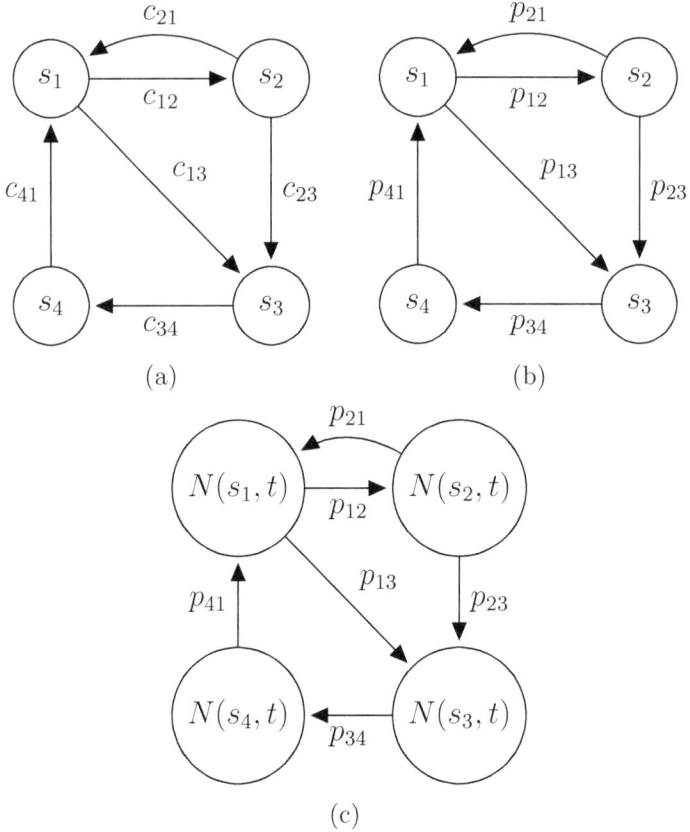

Figure 2.1: **(a)**: FSM representing a single robot controller with conditional edge transitions. Vertices represent physical or internal states that a robot can be in. **(b)**: PFSM of a robot controller with probabilistic edge transitions derived from simple geometric properties of the system. **(c)**: A macroscopic model for the swarm system as a whole. Vertices, $N(s_i, t)$, represent the number of robots in state s_i at time t. Edges are still transition probabilities between states.

equal to 5. This transition from deterministic FSM models to probabilistic PFSM models for swarm robot algorithms is derived in more detail in [13].

2.3 Mathematical Description of the System

Given a discrete set of states and conditions for transitions between them, usually in the form of probabilities of transition, a *master equation* defines a set of coupled ODEs that describe the time evolution of a physical system. So far, I have used logical constructs like FSMs to represent the robot controller running within each individual agent of the swarm system. I could instead look as these constructs as a model for the entire system, in which case the vertices of the FSM become accumulators of robots currently in a state and the edges define fractions of agents entering or leaving a given state at time t. The PFSM now becomes a macroscopic definition of the robot swarm and can be used to define a mathematical model for the time evolution of the system.

$$\frac{d\vec{P}(t)}{dt} = \mathbf{A}\vec{P}(t) \tag{2.1}$$

$\vec{P}$ is a vector containing the time-dependent probability of being in any given state in the corresponding PFSM. $\mathbf{A}$ is a matrix containing transition rates of going from state-i to state-j in the PFSM. When I multiply both sides of equation (2.1) by the total number of agents, N_0, I get the modified master equation that gives a macroscopic description of the system.

$$N_0\vec{P}'(t) = \mathbf{A}\left(N_0\vec{P}(t)\right)$$
$$\vec{S}'(t) = \mathbf{A}\vec{S}(t) \tag{2.2}$$

where $\vec{S}$ is a state vector containing the number of agents in each state, N_{s_i}, at time t. Here, $|\vec{S}|$ is equal to the number of unique states of the system, e.g. $|\vec{S}| = 4$ in my previous PFSM example from Figure 2.1b. The matrix $\mathbf{A}$ contains transition probabilities between the states in the PFSM. There a two types of elements, a_{ij} in matrix $\mathbf{A}$.

1. The non-diagonal entries, a_{ij} s.t. $i \neq j$, are equal to $p(c_{ij})$ (shortened to p_{ij}), the probability of transitioning from state s_i to s_j via the edge with conditional c_{ij} in the FSM.

2. The diagonal entries, a_{ii}, are equal to the negative sum of all edge probabilities p_n leaving state s_i.

If an edge does not exist between two states s_i, s_j ($i \neq j$) in the FSM, then entry $a_{ij} = 0$, e.g., the master equation for the swarm system described in Figure 2.1b is,

$$
\begin{pmatrix} N'_A(t) \\ N'_B(t) \\ N'_C(t) \\ N'_D(t) \end{pmatrix} = \begin{pmatrix} -(p_{12}+p_{13}) & p_{21} & 0 & p_{41} \\ p_{12} & -(p_{21}+p_{23}) & 0 & 0 \\ p_{13} & p_{23} & -p_{34} & 0 \\ 0 & 0 & p_{34} & -p_{41} \end{pmatrix}
\begin{pmatrix} N_A(t) \\ N_B(t) \\ N_C(t) \\ N_D(t) \end{pmatrix}
$$
(2.3)

In most of the scenarios being discussed in this paper, I assume that agents are neither removed nor added to a swarm system once an experiment has begun and therefore add the following constraints to the model,

$$
N_0 = \sum_{i=1}^{|\vec{S}|} N(s_i, t)
$$
(2.4)

$$
\forall j \leftarrow 1 \ldots |\vec{S}|, \sum_{i=1}^{|\vec{S}|} a_{ij} = 0
$$
(2.5)

Due to this constraint a simplification can be made to any one (but no more than one) of the states s_i in $\vec{S}$ so that,

$$
N(s_i, t) = N_0 - \sum_{j=1, j \neq i}^{|\vec{S}|} N(s_j, t)
$$
(2.6)

In swarm robotics literature, the master equation is often expanded to a set of difference equations or coupled ODEs called *rate equations* of the form,

$$
N'(s_i, t) = \sum_{j=1}^{|\vec{S}|} p_{ji} N(s_j, t) - \sum_{k=1}^{|\vec{S}|} p_{ik} N(s_i, t)
$$
(2.7)

along with a set of initial conditions that define the number of robots in each state at time 0. Rate equations are the preferred method for describing a macro-model of a swarm system because, unlike the master equation, they can represent probability values that could be complex, non-linear functions of environment variables, control variables, as well as time. These are also commonly referred to as population dynamics models.

2.4 Microscopic Simulation of the System

One of the advantages of using macroscopic, mathematical models for describing robot swarms is their ability to predict the state of the system at equilibrium, if it exists. But given the phenomenological approach to designing macro-models, it may not always be intuitive to construct the math equations to accurately describe the system. Even if the rate equations are defined, the system may not be easily solvable, either analytically or numerically. Fortunately there is another modeling tool that comes to our aid in such situations.

The Microscopic model (or micro-model) of a swarm system can be simulated using the *Gillespie* simulation technique [41, 42]. Here, each agent is simulated individually using dice rolls and probability. Gillespie developed this simulation algorithm in the 1970s to model the time evolution of reactant and product volumes in a chemical reaction. The individual agents in his chemical system were single molecules of the reactant and the micro-model was derived from the dynamics of molecule interactions. The probability of two reactant molecules colliding was computed using simple physical properties such as the radius and velocity of the molecules in the reaction medium [41]. Gillespie's original modeling approach has been modified for use with swarm systems, here. Martinoli outlines this process in detail in chapter 4.2 of his Ph.D. dissertation [7]. Robots with PFSM controllers are used instead of product and reactant molecules as the fundamental units of the simulation. First one step of the simulation, all robots are picked in a random order and their PFSMs are run in parallel for

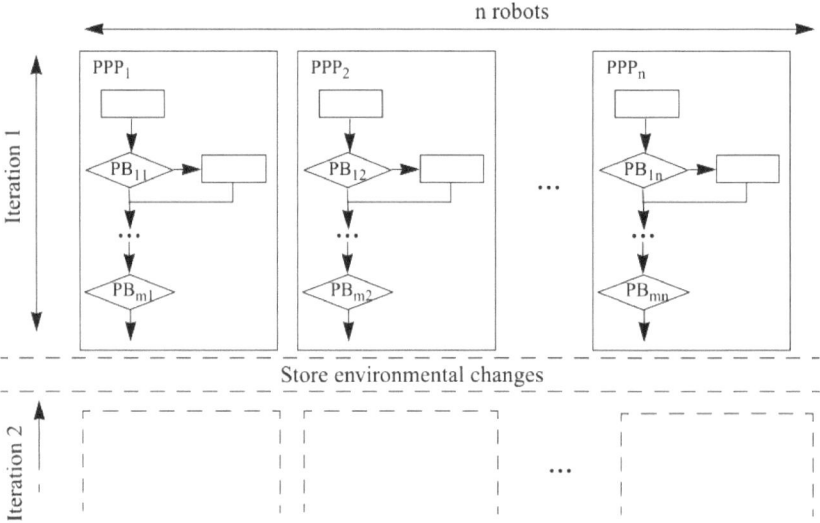

Figure 2.2: Gillespie simulation of a swarm system. Each controller, PPP_i, describes the independent behavior of a single robot in the swarm. Environmental variables are updated at the end of each iteration. (Image credit: Dr. Alcherio Martinoli)

a single time-step. The state of the entire system is then updated and the process repeats itself for a pre-determined length of time. Unlike in the original version of Gillespie simulation where time-step lengths between reactions are also chosen at random, here we preset the length of time that passes between two steps of the simulation.

2.5 Verification of System Properties Using Real Experiments and Physics-Based Simulation

An important step in any modeling process is validation by comparing model results to real experiment data. Given the relatively abstract approach applied so far for designing swarm robot models, this step is made even more crucial. The micro and macro-models in swarm robotics have conventionally been designed using

(a)   (b)

Figure 2.3: **(a)** The Webots simulator rendering a game of robot soccer. **(b)** The Droplet swarm robot simulator.

observed phenomena from other processes seen in biological and chemical systems and adapted to fit the swarming task being studied. Many swarm algorithms show emergent behavior where the observation of complex properties at the system level cannot be trivially inferred from studying the individual agent behavior. The generalizations and simplifications made in the robot controller design when developing the micro and macro-models can, and in many cases do, suppress the interesting emergent properties seen in real physical systems.

Physics-based simulators are often used to accurately recreate a task on a swarm system without investing the substantial time and resources required to develop and deploy real robots. These simulators attempt to remain as true to the real world as possible while maintaining an order of magnitude improvement in speed and simplicity over real robot experiments. Unlike micro-models that abstract away physical and environmental issues such as wheel slip, sensor noise, communication delays, etc., physics-based simulators make the added effort to accurately and dynamically model every minute aspect of the swarm system.

Many robot simulators are currently available today as either standalone programs like Webots (see Figure 2.3a). Webots is a widely used simulator in swarm robotics due to its capability to simulate multiple agents and agent-agent/agent-environment interactions in real time. The Webots API also allows for cross-compilation of programs from the simulation environment, right on to real robots without the need for reprogramming and supports a wide range of commercially available robot platforms such as Kilobots, Khepera, Alice, etc. There are also in-house implementations of physics simulators for specific robot platforms, such as the Droplet simulator shown in Figure 2.3b, that build up on physics engines such as Bullet and ODE.

2.6 Summary

The different methods discussed in this section are widely used for modeling and analyzing multi-agent systems and their corresponding algorithms. Each modeling methodology provides advantages at different levels of abstraction:

1. Gillespie simulation using FSM models allow us to study how micro-level individual agent interactions affect large scale behavior in the swarm.

2. PFSM models and rate equations allow us to study the dynamics of the system which provides insight on macro-level aspects such as steady state analysis, parameter identification, equilibria, etc.

3. Physics based simulations allow us to quickly verify our algorithms and provide a rapid prototyping tool for iterative refinement of swarm algorithms without having to invest in costly hardware.

The following chapter use these methodologies to define and analyze new task-assignment strategies for multi-agent systems.

Chapter 3

Designing an Optimal Control Model for Multi-Agent Systems

TA is a canonical problem in swarm robotics. With the advent of more sophisticated general purpose robots such as the Baxter, NAO and innumerable UAVs the problem of handling real world tasks collaboratively, be it with other robots or humans, is becoming increasingly important. Many group tasks exhibit a property of concurrent benefit, i.e. single agent attempts to complete the task are guaranteed to fail or waste resources but groups attempting the task together provide a considerable concurrency benefit. A problem oft ignored by MATA model developers is deciding when a group is capable enough to attempt the collaborative task and whether or not simultaneous actions are important to complete the task successfully. While this precise temporal component of attempting tasks concurrently is not the focus of our work, it is important to point out as we make certain assumptions on required task group sizes and how these requirements change with time. In this context TA is distilled down to the process of assigning the required number of agents to particular dynamic tasks without worrying about how agents get to the tasks or what the exact dynamics of agents and these genericized tasks are.

We assume that tasks or targets (used synonymously throughout this paper) require at least two agents to attempt collaboratively. The major caveat here is that the *exact* number of agents required to attempt a task is unknown and very difficult to accurately discern. Tasks with concurrent benefit share the property that the probability of success depends non-linearly on the collective capabilities and team size of the robots attempting it. In addition, the exact number of agents required to successfully complete the task varies over time due to numerous complex physical parameters. Many collaborative tasks—particularly those seen in biological systems—exhibit the property of concurrent benefit, ranging from surveillance and coordinated defense of enclosed areas like termite mounds and honey bee hives [43] to collective transport of heavy objects and even containment of oil spills and forest fires. To contain a large fire, it is insufficient (and inefficient) for a single agent to start putting out the fire without waiting for backup. But the rate of fire containment increases quickly by adding just a few more agents to the group, which illustrates the property of concurrent benefit well.

3.1 Multi-Agent Task Allocation Model

With this general overview of MATA in mind, we present the following formal model for TA.

- Agents/Robots: $\mathcal{P} = \{n_1, n_2, \ldots, n_i, \ldots, n_{|\mathcal{P}|}\}$

- Targets: $\mathcal{T} = \{t_1, t_2, \ldots, t_j, \ldots, t_{|\mathcal{T}|}\}$

- Target Threshold: $K : \mathcal{T} \to \mathbb{Z}^+$
  The number of agents required to successfully attempt task-t_j is $= K(t_j)$, which is shortened to k_j for brevity.

- Agent Constraints: $C : \mathcal{P} \to \hat{\mathcal{T}} \subseteq \mathcal{T}$
  The set of constraints for agent-n_i ($= C(n_i)$) is the subset of targets that this agent can reach. It is shortened to c_i for brevity.

- Agent Assignment Matrix: A $|\mathcal{P}| \times |\mathcal{T}|$ matrix of 0-1 elements $\xi(n_i, t_j)$ or ξ_{ij} for short, that are either 0 if agent-i is not assigned to target-j or 1 if agent-i is assigned to target-j.

$$\chi = \begin{pmatrix} \xi_{11} & \cdots & \xi_{1|\mathcal{T}|} \\ \vdots & \ddots & \vdots \\ \xi_{|\mathcal{P}|1} & \cdots & \xi_{|\mathcal{P}||\mathcal{T}|} \end{pmatrix} \tag{3.1}$$

- Target Assignments: $\mathcal{A}: \mathcal{T} \to \hat{\mathcal{P}} \subseteq \mathcal{P}$
  The set of agents assigned to target-t_j is $= \mathcal{A}(t_j)$, which is shortened to α_j for brevity. From (3.1) we can define

$$|\alpha_j| = \sum_{i=1}^{|\mathcal{P}|} \xi_{ij} \tag{3.2}$$

Definition: A target is considered "successfully assigned" when $|\alpha_j| \geq k_j$, i.e. the number of player's assigned to it is greater than or equal to its threshold value.
Definition: A target is considered "perfectly assigned" when $|\alpha_j| = k_j$.

- Target specific welfare function,

$$W(t_j, |\alpha_j|) = \begin{cases} w_j & |\alpha_j| \geq k_j \\ 0 & o/w \end{cases} \tag{3.3}$$

where w_j can be a value or function defining the utility of completing task-t_j upon successful assignment.

- Global welfare function,

$$\text{Maximize} \quad \mathcal{W} = \sum_{t \in \mathcal{T}} W(t, |\mathcal{A}(t)|) \tag{3.4}$$

$$\text{S. T. for all} \quad i = 1 \ldots |\mathcal{P}|, j = 1 \ldots |\mathcal{T}|,$$

$$\sum_{t \in \mathcal{T}} \xi(n_i, t) \leq 1$$

$$\sum_{t \in \mathcal{T} \backslash C(n_i)} \xi(n_i, t) = 0$$

$$0 \leq \xi(n_i, t_j) \leq 1$$

Eq. (3.4) is an ILP that can be maximized under constraints to provide an *optimal* assignment of agents to targets. This is in contrast to a lot of existing approaches for MATA where each agent is concerned with maximizing their own utility, making such approaches agent-centric. We, instead, focus on the task as the primary entity for which a utility function is defined and maximized. This approach falls more in line with work such as [44]. The constraints listed in Eq. (3.4) ensure that an agent cannot be assigned to more than one target and that an agent can never be assigned to targets outside of its constrained target set $C(n_i)$. The final constraint just ensures that $\xi(n_i, t_j)$ is only ever 0 or 1 since this is a 0-1 ILP.

The following example describes a cooperative game with target thresholds and player constraints as seen in Figure 3.1.

- Agent: $\mathcal{P} = \{1, 2, 3, 4\}$

- Targets: $t \in \mathcal{T} = \{a, b, c\}$

- Target thresholds: $k_a = k_b = k_c = 2$

- Agent constraints: $c_1 = c_3 = \{a, b\}$ and $c_2 = c_4 = \{b, c\}$

- Target specific welfare function: $W(t_j, |a_j|) = 1$ (if $|a_j| \geq k_j$), 0 otherwise.

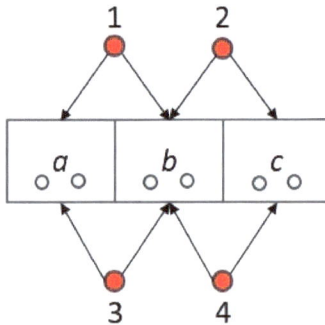

Figure 3.1: A cooperative game with 4 players and 3 targets. Gray pegs indicate the target's minimum threshold value while the arrows depict player assignment constraints.

3.2 Defining Optimality for Multi-Agent Task Allocation

Consider a simplified forest firefighting scenario where a number of isolated fires and flame fronts exist in a specified geographic region. Our goal is to contain all of the fires and prevent them from spreading or, at the very least, contain as many as we can given our manpower. If the fires did not evolve over time and we had perfect information about the threshold value of each target/fire then a central controller would only have to maximize Eq. (3.4) and provide the resulting assignment matrix to all the agents. The agents would then move to their assigned fires. Since Eq. (3.4) guarantees successful assignments (if such an assignment is possible) there would be enough robots at each fire to put it out and all the tasks would be complete.

Clearly this situation — like most real world situations — is dynamic, i.e., the size and number of fires changes over time. The task allocation model described in the previous section has no notion of time. We consider constant but repeated optimization of *snapshots* of a dynamic problem so that at any given point a solution to the TA problem is valid. The main assumption made here is that maximizing Eq. (3.4) takes considerably less time than the physical evolution of the task magnitudes, and in-turn, the task thresholds. If this is not the case or if, for any other reason, the state of the system is not updated in a central controller then this notion of optimality quickly fails.

Nevertheless, maximizing snapshots of a MATA system with perfect information throughout the course of a prescribed scenario provides a means for defining baseline system-wide optimality. At any given point in time, the optimal solution of a snapshot of the state of the system is the best assignment of targets that agents can be given.

Chapter 4

Existence of an Equilibrium Strategy for Communication Free Multi-Robot Task Assignment

A number of different methods exist for TA in MAS such as deterministic leader-follower coalition algorithms [18] and more complex market-based approaches [16] to simpler probabilistic algorithms for individually simplistic agents [19]. As mentioned earlier, the reason that we focus so heavily on the RT model is it's widespread use in ethology research for modeling social insects. While this phenomenological evidence provides a solid incentive for their use in engineered robotic systems, there has so far been no mathematical argument for their use. In this chapter we provide — for what we believe to be the first time in the field of swarm robotics — a mathematical explanation for their prevalence using a well known result from game theory. The problem of TA in MAS is formulated as a global game and two theorems are

I would like to thank Dr. Behrouz Touri for his extensive help on the theory of global games and in formulating and proving the theorems mentioned in this chapter. The use of the article "we" in this chapter serves to remind the reader of the extent our my collaboration with Dr. Touri.

subsequently proved showing the existence of a Bayes Nash Equilibrium in a system using DRTs (Theorem 1) and, subsequently, CRTs (Theorem 2) for TA in MAS. Theorem 2 also analytically shows how Gaussian noise in perceiving stimulus signals leads to sigmoid RTs, as are commonly observed in nature.

4.1   Global Games: A Brief Overview

Game theory is the study of strategic interactions among multiple agents or players, such as robots, people, firms, etc. where the decision of each party affects the payoff of the rest. A fundamentally important class of games is one with incomplete or imperfect information where each agent's utility depend not only on the actions of the other agents, but also on an underlying fundamental signal that cannot be accurately ordained by the agents. The class of global games with incomplete information was originally introduced in [45] where two players are playing a game and the utility of the two players depends on an underlying fundamental signal $\tau \in \mathbb{R}$, but each agent observes a noisy variation of this signal, x_i. For example in a firefighting task, this fundamental signal τ is the *magnitude* of the task of putting out the fire, i.e., the number of robots needed to do so. The size and intensity of the fire, along with environmental and other site-specific factors all play a major role in determining whether an agent should begin the task or wait for more help to arrive.

    While we use the term magnitude to describe τ, it is a standin for a simplified representation of a more abstract quality of any task. All tasks demand completion and the act of completion requires resources, be it time and/or energy of some form. In swarms of minimalist agents with limited capabilities, the resource required to collaboratively complete a task is invariably quantized into the number of agents attempting to complete that task. In bee hives and ant colonies the drive to complete a task is regulated by pheromone levels, among other signals. In engineered swarm systems, more direct measurements of the environment through the use of on-board sensors allow a robot to independently estimate τ, however imperfectly. In either case, τ is an inherent *truth* about the task that can never be discerned accurately, but is always indirectly estimated by all agents. It is important to highlight that agents never share their independent estimates of τ with each other in my proposed formulation of the TA problem. While this may seem like a weakness in my argument, most of the research on RT TA explicitly mentions limited to no communication requirements to be a major advantage of this approach versus other methods since even limited information propagation through a > 1000 agent system quickly becomes the bottleneck for any distributed swarm algorithm.

4.2   Task Allocation as Global Games

Consider a group of agents performing a task contributing to a common goal, which we refer to as a concurrent benefit. This benefit is related to a stimulus τ that can be observed by all agents, albeit subject to sensing noise. Agents do not share any information. All agents decide, for themselves, whether or not to engage in the task. A task is successfully attempted if a critical mass of agents is willing to participate in it. Otherwise, the attempt fails.

Situations like this arise in a number of different fields including neurology [46, 47], ethology [27, 30, 48, 49], sociology [50], economics [51], and robotics [12, 15, 34, 52–56]. All of these multi-agent scenarios share the common notion of a joint action or response to a commonly observed stimulus. The task can take on many forms ranging from neurons simply firing in concert, collective decision problems like flocking, herd grazing and colony defense to individual actions based on the environment and other agents' beliefs like foraging, bank runs, and political revolutions.

In the case of a bank run [51], τ is an aggregate stimulus parameter that represents the strength of the economy of a nation. Here, agents decide when to withdraw their assets from banks based on their own noisy estimate of the economy together with a simple threshold. In the case of social insects foraging for food [25, 30, 34], τ represents a number of environmental cues such as the (imperfect) measurement of food stores in a colony, pheromone levels [27] or the waiting time for food transfer from one agent to another [57].

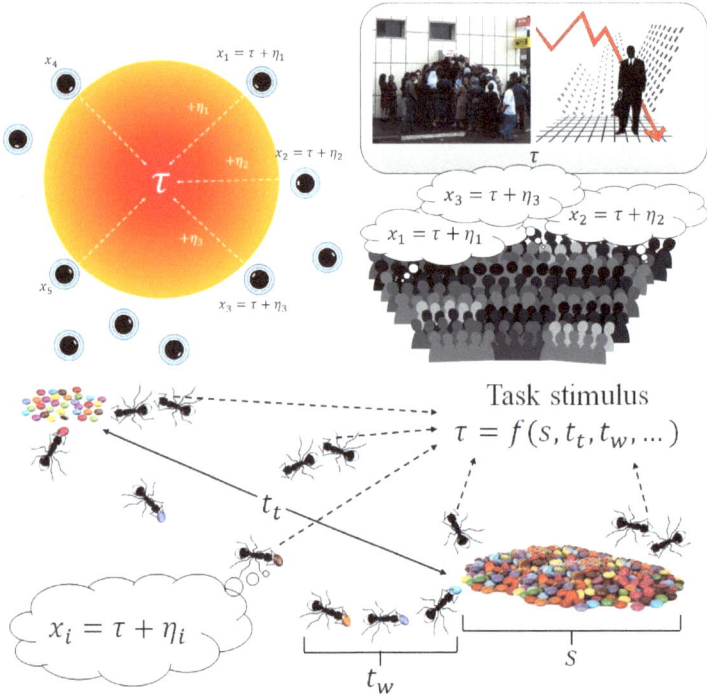

Figure 4.1: Robotic fire fighting, ant foraging, and bank run scenarios presented as global games. Each player's imperfect estimate of the task is represented by x_i, comprising of the global stimulus parameter τ and noisy sensor measurements η_i. In the robot firefighting scenario τ is representative of the magnitude of the fire, while in the case of a bank run τ is indicative of an agent's current level of trust in the nation's economy. For the ant foraging scenario τ represents an ant's willingness to take part in the foraging task based on a number of internally measured parameters such as the distance to the food source (t_t), the wait time to deliver food (t_w), and the food stores currently at the nest (s), among others.

A complex combination of these internal and external cues [48] temper an agent's perception of the magnitude of a task. In an engineering context, τ can be seen as the magnitude of a fire (heat intensity and area covered) as sensed by a robot using on-board instruments in an automated firefighting scenario [12]. Figure 4.1 illustrates each of these three examples with their corresponding stimulus parameters.

The group dynamic in the above examples may seem orthogonal at first; while adversarial behaviour between agents drives bank runs, collaborative behaviour between robots is essential for the automated firefighting scenario. Both scenarios, however, share the notion that to be successful an agent not only needs to assess the magnitude of the task itself but also the likelihood of the other agents to act. This is because only acting in concert leads to the desired group action, be it because using up water resources to put out a fire is futile before critical mass is reached, or disengaging from the banking system is non-desirable unless there is a major crisis. In a system with multiple tasks, such as an ant colony, coordination is required to achieve a desirable proportion between tasks.

We build on results from global games [45] to show that the observed behaviour in all these scenarios can be effectively emulated by assuming that each agent makes their individual decision on whether or not to perform a task based on some internal threshold value which is compared to their noisy estimates of the collective task's stimulus τ. This was shown for the canonical bank run example [51]. While the classical global game assumes each agent must predict the other agents' behaviour, it turns out that agents can reach an equilibrium without this capacity. This, and the fact that agents do not need to communicate, makes this approach widely applicable to a wide range of multi-agent systems.

Consider a set of n agents and suppose that each agent has an action set $A_i = \{0, 1\}$ where 0 represents not participating in the task and 1 represents participating in the task. Every agent is also aware of the total number of other agents, n in the system. For the purpose of analysis we assume the decision to act or not to act is made by all agents at the same time, i.e. this is a one-shot game with no notion of time. We let the stimulus τ be a real number that belongs within the interval $E = [c, d]$ in $\mathbb{R}$. Finally, we let $u_i : A_i \times \mathbb{Z}^+ \times \mathbb{R} \to \mathbb{R}$ be the utility of the ith agent, where $u_i(a_i, g, \tau)$ is the utility of the ith agent when g other agents have decided to participate in the task. In general, the utility of each agent depends on the joint actions of the rest of the agents. For simplicity, we assume the utility to be proportional to the number of agents participating in the activity.

The utility function discussed throughout this paper has the following properties:

a. $u_i(1, g, \tau) - u_i(0, g, \tau)$ is an increasing and continuous function of τ for any g. We further assume that $|u_i(1, g, \tau) - u_i(0, g, \tau)| \leq \tau^p$ for some $p \geq 1$.

b. For extreme stimulus ranges, taking part in the activity is either appealing or repelling, i.e. there exists $\underline{\tau}, \bar{\tau} \in (c, d)$ with $\underline{\tau} \leq \bar{\tau}$ such that for any $\tau \geq \bar{\tau}$ we have $u_i(1, g, \tau) > u_i(0, g, \tau)$ where all agents participate in very easy tasks, and for $\tau \leq \underline{\tau}$ we have $u_i(1, g, \tau) < u_i(0, g, \tau)$ so the only equilibrium of the game is for all agents to not participate as the task is too difficult.

Note that in order to have a task with such an utility, we need the above conditions to hold for all the agents, i.e. for all $i \in \{1, \ldots, n\}$. An example of a utility function that would satisfy such conditions is a function $u_i(a_i, g, \tau) = a_i(1 - e^{-(g+1)} + \tau)$.

The main challenge in devising task allocation strategies is that the true value of τ is not easily accessible to the agents, for example due to limited perception capabilities and sensor noise. We model this imperfect knowledge by assuming that agent i observes $x_i = \tau + \eta_i$ where η_i is a Gaussian $\mathcal{N}(0, \sigma_i^2)$ random variable. Note that this makes the game a Bayesian game and in this case, the type of each player is represented by the random variable x_i. Throughout our discussion, we assume that the task stimulus τ is a Gaussian random variable and is independent of $\eta_1, \ldots, \eta_n$. This analysis is extendable to a larger class of random variables but for the simplicity of the discussion, we consider Gaussian random variables here. Given these constraints, the question is what strategy the agents should follow to reach a BNE. In other words, an outcome in which no agent has the incentive to deviate from its current strategy.

A *strategy* s_i for the ith agent is a measurable function $s_i : \mathbb{R} \to A_i$, mapping measurements (observations) to actions. Strategy s_i prescribes what action the ith agent should take given its own measurement (type) x_i. Given this, consider a set of agents with strategies $s_1, \ldots, s_n$. Let us denote the strategies of the $n - 1$ agents other than the ith agent by the vector $S_{-i} = \{s_1, \ldots, s_{i-1}, s_{i+1}, \ldots, s_n\}$. We say that a strategy s_i is a *threshold strategy* if $s_i(x) = \text{step}(x, \Upsilon_i)$, i.e. the step function with a jump from 0 to 1 at Υ_i, where Υ_i is the internal threshold value of the ith agent. For the ith agent, we define the best-response $BR(S_{-i})$

(to the strategies of the other agents) to be a strategy $\tilde{s}$ that for any $x \in \mathbb{R}$:

$$BR(S_{-i})(x) = \tilde{s}(x) \in \arg\max_{a_i \in A_i} E(u_i(a_i, g, \tau) \mid x_i = x) \qquad (4.1)$$

$$= \arg\max_{a_i \in A_i} E(u_i(a_i, \sum_{j \neq i} s_j(x_j), \tau) \mid x_i = x), \qquad (4.2)$$

where $E(\cdot|\cdot)$ is the conditional expectation of u_i given the ith agent's observation. The best response of player i simply is the best course of action for agent i given that the strategies of the other players is given. Indeed the expression $\arg\max_{a_i \in A_i} E(u_i(a_i, g, \tau) \mid x_i = x)$ is the (set of) best actions that player i can take given its information, the aggregate action g of the other players, and the intensity τ. Note that given the ith agent's observation x_i, the observations of the other agents, and hence their actions, would be random from the ith agent perspective, i.e. given x_i and τ, all $s \in S_{-i}$ are effectively random variables with respect to the ith agent. A strategy profile $S = \{s_1, \ldots, s_n\}$ is a *sensible strategy*, if it leads to a BNE [58], given $s_i = BR(S_{-i})$ for all $i \in \{1, \ldots, n\}$.

4.3 Communication Free Threshold Based Task Allocation Strategy

Any task with concurrent benefit admits a threshold strategy BNE — meaning it is sufficient for the agents to follow a simple algorithm:

(i) Compare your noisy measurement x_i to a threshold value Υ_i,

(ii) If the measurement is above Υ_i take part in the collaborative task, otherwise hold off.

This algorithm is extremely simple, and can be implemented on systems with a wide range of capabilities, yet leads to a BNE as we will show below.

To show that there exists a sensible threshold strategy for the class of tasks with concurrent benefit leading to Theorem 1, we will first show that the best response to threshold strategies is a threshold strategy (Lemma 1), and then show that there exists an equilibrium of threshold strategies [45, 51] (Lemma 2).

Lemma 1. *Let* $S = \{s_1, \ldots, s_n\}$ *be a strategy profile consisting of threshold strategies for a task with concurrent benefit. Let* $\tilde{s}_i = BR(S_{-i})$. *Then* $\tilde{s}_i$ *is a threshold strategy.*

Proof. We first show that if for some observation $x_i = x$, we have $BR(S_{-i})(x) = \tilde{s}_i(x) = 1$, then $\tilde{s}_i(y) = 1$ for $y \geq x$. To show this, we note that $P(x_j \geq \tau_j \mid x_i = x)$ is an increasing function of x as $x_j - x_i$ is a normally distributed random variable. Therefore, using the monotone property of concurrent tasks and the fact that $x_i = \tau + \eta_i$, we conclude that:

$$E(u_i(1, \sum_{j \neq i} s_j(x_j), \tau) \mid x_i = y) \tag{4.3}$$

$$- E(u_i(0, \sum_{j \neq i} s_j(x_j), \tau) \mid x_i = y) \tag{4.4}$$

$$> E(u_i(1, \sum_{j \neq i} s_j(x_j), \tau) \mid x_i = x) \tag{4.5}$$

$$- E(u_i(0, \sum_{j \neq i} s_j(x_j), \tau) \mid x_i = x) \geq 0. \tag{4.6}$$

Therefore $\tilde{s}_i(y) = 1$. Similarly, if for some value of x, we have $\tilde{s}_i(x) = 0$, then it follows that $\tilde{s}_i(y) = 0$ for $y \leq x$. Therefore, $\tilde{s}_i$ would be a threshold strategy. $\square$

We can view the best-response of threshold strategies as a mapping from $\mathbb{R}^n$ to $\mathbb{R}^n$ that maps n thresholds of the original strategies to n thresholds of the best-response strategies. Denote this mapping by $L : \mathbb{R}^n \to \mathbb{R}^n$.

Lemma 2. *The mapping L that maps the threshold values of threshold strategies to the threshold values of the best-response strategies is a continuous mapping.*

Proof. Let $x_{-i} = (x_1, \ldots, x_{i-1}, x_{i+1}, \ldots, x_n)$ be the vector of observations of $n-1$ agents except the ith agent. Note that the vector (x_{-i}, τ) given $x_i = x$ is a normally distributed random vector with some continuous density function $f_x(x_{-i}, \tau)$. Now, let $\{\alpha(k)\}$ be a sequence in $\mathbb{R}^n$ that is converging to $\alpha \in \mathbb{R}^n$. Let $\{\beta(k)\}$ be the sequence of thresholds corresponding to the best-response strategy of the strategy with threshold vector $\alpha(k)$. Let s be the threshold strategy corresponding to the threshold vector α and let α^* be the threshold strategy corresponding to the $BR(\alpha)$. By the definition of the best-response strategy, $\beta_i(k)$ is a point where

$$\int_{\mathbb{R}^n} f_{\beta(k)}(z, t) \left(u_i(1, \sum_{j \neq i} u^{\alpha_j(k)}(x_j), \tau) \right. \tag{4.7}$$

$$\left. -u_i(0, \sum_{j \neq i} u^{\alpha_j(k)}(x_j), \tau) \right) d(z \times t) = 0. \tag{4.8}$$

Using the fact that f has a Gaussian distribution and is continuous on all its arguments and the fact that $|u_i(\cdot, \cdot, \tau)| \leq \tau^p$, by taking the limit $k \to \infty$ and the dominated convergence theorem:

$$\int_{\mathbb{R}^n} f_\beta(z, t)(u_i(1, \sum_{j \neq i} u^{\alpha_j}(x_j), \tau) \tag{4.9}$$

$$- u_i(0, \sum_{j \neq i} u^{\alpha(k)}(x_j), \tau))d(z \times t) = 0, \tag{4.10}$$

where u^r is a threshold strategy with threshold r. Therefore, the $\lim_{k \to \infty} L(\alpha(k)) = L(\alpha)$ for a sequence $\{\alpha(k)\}$ that is converging to α. □

Using these lemmas, we can show the existence of a threshold strategy for global games with concurrent benefit.

Theorem 1. *For a concurrent benefit task T, suppose that the stimulus parameter τ is a Gaussian random variable. Also, suppose that $x_i = \tau + \eta_i$ where $\eta_1, \ldots, \eta_n$ are independent Gaussian random variables. Then, there exists a strategy profile $S = (s_1, \ldots, s_n)$ of threshold strategies that is a BNE.*

Proof. By Lemma 1, the best response of a threshold strategy is a threshold strategy and hence, it induces the mapping L from the space of thresholds $\mathbb{R}^n$ to itself. Also, by Lemma 2, this mapping is a continuous mapping. Now, if Υ_i is a sufficiently large threshold,

then the second property of concurrent benefit tasks implies that
the $\tilde{\Upsilon}_i \leq \Upsilon_i$ because a large enough measurement x_i implies that
agent i itself should take part in the task. Similarly, for sufficiently
low threshold Υ_i, we will have $\tilde{\Upsilon}_i \geq \Upsilon_i$. Therefore, the mapping
L maps a box $[a, b]^n$ to itself, where a is a sufficiently small scalar
and $b > a$ is a sufficiently large scalar. Since the box $[a, b]^n$ is
a convex closed set, by the Brouwer's fix point theorem [59] we
have that there exists a vector of threshold values α^* such that
$\alpha^* = L(\alpha^*)$ and hence, there exists a BNE for the concurrent
benefit task T. □

4.4 From Discrete Thresholds to Sigmoidal Response Functions

Observations in ethology suggest sigmoid threshold functions [25],
rather than fixed thresholds as suggested by our analysis. Also,
roboticists have started using sigmoid-shaped threshold functions
to engineer swarm systems [25, 30, 34], as tuning the shape of a
sigmoidal response threshold function allows balancing between
exploration, i.e., performing a random action, and exploitation,
i.e., using all available information in decision making such as a
fixed threshold. We argue that this behavior can be a direct result of using a simple discrete threshold under the influence of
perception noise. Indeed, one can show that a sigmoid threshold function is the outcome of deterministic threshold functions
on noisy observations. Suppose that all agents share the same
utility function $u(a_i, g, \tau)$ and also, assume that the observation
noise of the n agents $(\eta_1, \ldots, \eta_n)$ are independent and identically
distributed (IID) $\mathcal{N}(0, \sigma^2)$ Gaussian random variables. Then, it is
not hard to see that there exists a BNE with threshold strategies
that have the same threshold value Υ [51].

Consider a realization of $\tau = \hat{\tau}$ and suppose that we have a
large number of agents n observing a noisy variation of $\hat{\tau}$. Take for
example the case of fire-fighting agents, and let $\hat{\tau}$ be the magnitude
(including type, intensity, area, etc.) of the fire. Then, since the
observations of the n agents are IID given the value of τ, they will

be distributed according to $\mathcal{N}(\hat{\tau}, \sigma^2)$. Now consider the relative number of agents taking part in the activity given $\hat{\tau}$ as defined by,

$$N_{rel}(\hat{\tau}) := \frac{\#\text{agents with } x_i \geq \Upsilon}{n}.$$

We can now show that for the relative number of agents $N_{rel}(\hat{\tau})$, we have

$$\lim_{n \to \infty} N_{rel}(\hat{\tau}) = \Phi(\frac{\hat{\tau} - \Upsilon}{\sigma^2}) \tag{4.11}$$

where Φ is the cumulative distribution function (cdf) of a standard Gaussian, which is illustrated numerically in Figure 2, and shown in Theorem 2:

Theorem 2. *For the relative number of agents $N_{rel}(\hat{\tau})$, we have*

$$\lim_{n \to \infty} N_{rel}(\hat{\tau}) = \Phi(\frac{\hat{\tau} - \Upsilon}{\sigma^2}) \tag{4.12}$$

where Φ is the cumulative distribution function (cdf) of a standard Gaussian.

Proof. Note that $N_{rel}(\hat{\tau}) = \frac{\sum_{i=1}^{n} \mathbf{I}_{x_i \geq \Upsilon}}{n}$ where $\mathbf{I}_{i \geq j}$ is the indicator function for $i \geq j$. For a given $\hat{\tau}$, x_i are IID $\mathcal{N}(\hat{\tau}, \sigma^2)$ random variables and hence, $\mathbf{I}_{x_i \geq \Upsilon}$ are IID random variables for all agents with $E(\mathbf{I}_{x_i \geq \Upsilon}) = \Phi(\frac{\hat{\tau} - \Upsilon}{\sigma^2})$. Therefore, by the Law of Large Numbers, it follows that:

$$\lim_{n \to \infty} N_{rel}(\hat{\tau}) = \Phi(\frac{\hat{\tau} - \Upsilon}{\sigma^2}). \tag{4.13}$$

$\square$

The final step to explain the prevalence of sigmoid functions in multi-agent settings is to note that:

$$\left|\Phi(\frac{\hat{\tau} - \Upsilon}{\sigma^2}) - \frac{1}{1 + e^{-d(\frac{\hat{\tau} - \Upsilon}{\sigma^2})}}\right| \leq 0.01, \tag{4.14}$$

for all $\hat{\tau} \in \mathbb{R}$ and some optimal value $d \approx 1.704$ as described in [60]. This means that the aggregate behavior of the agents following deterministic threshold strategies would closely follow (to within a constant error term) the shape of the commonly observed logistic sigmoid function whose drift is directly proportional to Υ and the slope is inversely proportional to σ^2.

Therefore, despite agents using deterministic threshold strategies, their *aggregate behavior* would appear to an outside observer as a continuous sigmoid threshold function instead of numerous discrete thresholds.

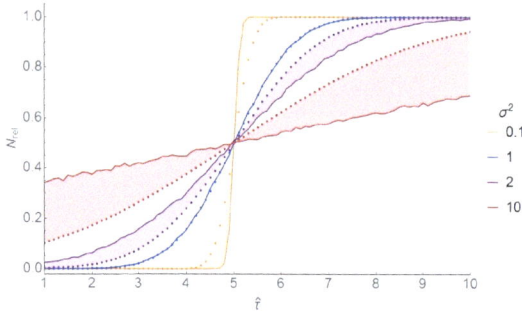

Figure 4.2: Visualization of Theorem 2 as N_{rel} estimates $\Phi(\cdot)$. The plot was generated by running Eq. 4.12 10,000 times for each point in $\hat{\tau} = 1$ to 10 in increments of 0.1. $n = 10$, $\Upsilon = 5$ and $x_i = \hat{\tau} + \eta_i$ ($\eta_i \sim \mathcal{N}(0, \sigma^2)$). Each solid-line in the plot is generated by sweeping $\sigma^2 = \{0.1, 1, 2, 10\}$, with $\sigma^2 = 0.1$ being close to a step-function and $\sigma^2 = 10$ having the *flattest* slope. The shaded region provides a difference comparison between the N_{rel} estimate of $\Phi(\cdot)$ and $\Phi(\cdot)$ itself, which is plotted using dotted-lines.

4.5 Discussion and Summary

I show in Theorem 1 that a communication-free agent-level threshold strategy is sufficient and necessary to achieve a TA resulting in system-level equilibrium for concurrent benefit tasks. I then show in Theorem 2 how such a policy effectively reduces to a continuous threshold response function that is commonly observed in social insects. While this result explains why one would observe such phenomena in natural systems, the game theory perspective leaves it unclear why one should choose a continuous threshold function in a robotic setting. Here, recall that a benefit of using CRT vs. DRT is the randomness it adds, allowing swarm systems to quickly adapt to changes in task and environment parameters [26]. In particular, deliberately tuning the slope of the sigmoid threshold function as I have investigated in [12] allows a swarm to explore different team sizes, and possibly learn from this experience, an aspect I wish to study in future work.

Theorem 1 states that there exists a unique strategy profile s such that it is a system-level Bayesian Nash Equilibrium for all agents performing a concurrent benefit task. Notice that this theorem makes no claims towards an *optimal* strategy for TA, just an equilibrium strategy. It is important to distinguish between these two properties of the swarm system. Clearly, optimal outcomes will require communication among agents or, at least, a central entity with access to global information. In in the next chapter I the area in-between where agents exchange limited amounts of information, such as their noisy estimates x_i.

This raises the question what "communication" and "sharing of information" actually mean. In this chapter, communication is limited to common observations of an unknown environmental variable τ. Natural swarming systems often communicate by modifying the environment, a form of indirect communication known as stigmergy [37]. The key difference between this form of communication and direct exchange of information, as I discuss in [61], is that the same information is simultaneously accessible to all individuals. I therefore believe that the results presented here extend also to indirect communication via stigmergy where τ is the measurable result of previous agent activity.

Chapter 5

Response Threshold Model for Multi-Agent Task Allocation

Drawing inspiration from TA in social insects, this chapter describes a novel approach to recruiting a variable number of agents for a particular task. Using CRT functions I show that the resulting macro-level agent team-size mean and variance can be controlled within desired ranges using only two micro-level parameters. Controlling variable team-sizes for TA in a MAS is an important step towards optimal control of the system as it uniquely identifies the important parameters involved in group size estimation of tasks of varying magnitude and difficulty.

I consider a generic collaboration task with m uniformly distributed collaboration sites within a flat arena with area A. A swarm of individually simple robots such as the *Droplet* platform [62, 63] is deployed within the arena, uniformly and at random. The number of robots being used per experiment varies, as I discuss results for a number of different scenarios. Collaboration sites in the arena can be of various sizes and configurations.

Each individual agent is capable of locomotion [63] and local sensing [62]. The agents do not require global positioning and no centralized controller exists, but I assume each agent to be capable of local omnidirectional communication with other agents within its communication range. The agents are also capable of sensing the boundary of a collaboration site—I assume that sites have easily distinguishable boundary regions, as shown in Fig. 1.1, for the purposes of the model studied in this paper.

The objective of each agent in the robot swarm is to find a collaboration site in the arena and perform a collective task with other agents at that site. The precise details of the collective task are not important for the purpose of understanding the coordination mechanism. I assume the actual collective task takes each agent a probabilistic finite amount of time to complete. Once collaboration is complete, the agent detaches itself from its current site and returns to searching for other sites in the arena.

It is perfectly reasonable to assume that agents arrive at the same collaboration site after having just completed a task there (possibly unsuccessfully) but will now be part of a new collaboration group. Each agent individually decides whether or not to collaborate at a given time step, while waiting at a collaboration site. If the majority of agents at that site decide to collaborate then the entire population is recruited for the task and thus a collective consensus is reached using a majority voting scheme. Here, I consider a majority to mean exactly half or more of a given population.

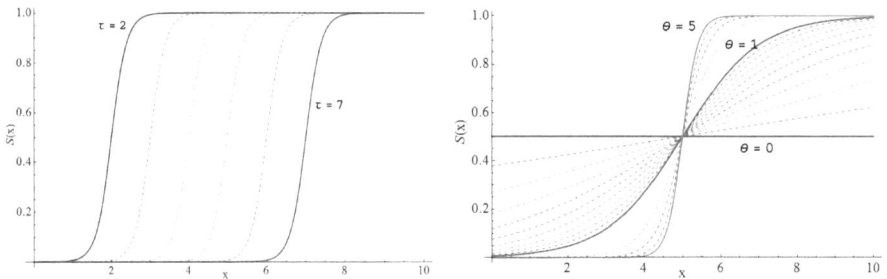

(a) Changing τ offsets the curve along the x axis, allowing to set the desired mean team size.

(b) Changing θ changes the slope at the point $x^* = \tau$, $\mathcal{S}(x^*) = 0.5$, allowing to control the team's variance.

Figure 5.1: Sigmoid CRT function and its parameters.

An individual agent-i's willingness to collaborate is a stochastic term governed by a sigmoid based RT function that takes as input, the number of agents $x_{\hat{m}}$ currently at the same collaboration site as agent-i and outputs a probability of collaboration using control

Chapter 5. Response Threshold Model

parameters θ and τ:

$$S(x_{\hat{m}}) = \frac{1}{1 + e^{\theta(\tau - x_{\hat{m}})}} \tag{5.1}$$

The parameter θ controls the slope of the sigmoid function, while τ controls its offset along the x axis, as seen in Fig. 5.1. Each agent is independently responsible for estimating the group size $x_{\hat{m}}$ at a given time either by direct sensing or by communication. In practice, this involves building a list of unique identifiers of the agents sharing its collaboration site. The overall algorithm, followed by each individual agent in the system, is provided in Alg. 1.

Note that the proposed RT function is different from [64], who uses high-order polynomials. While these functions work well in regimes with moderate slope, they create numerical problems when approximating unit-step-like responses such as those (implicitly) used in [20]. I particularly chose the logistic function from the large class of sigmoid functions due to the intuitive nature of the parameters τ and θ.

Algorithm 1 TA algorithm for an individual agent using the sigmoid threshold function

---

  function TASK_ALLOCATION(θ, τ)
    $estimate \leftarrow$ discover_group_size()
    $decision \leftarrow$ run_sigmoid($estimate$, θ, τ)
    communicate_decision($decision$)
    $decisions[] \leftarrow$ gather_decisions()
    $result \leftarrow$ Count($decisions[]$, $true$)   ▷ Count() Returns the number of successes in the decisions
    if $result \geq (estimate/2)$ **then**
      Collaborate()
      return
    else
      Task_Allocation(θ, τ)
    end if
  end function

5.1 Macroscopic analysis

In this section I study how the local parameters τ and θ from an individual agent's sigmoid threshold function affect formation of groups of different sizes at the macroscopic system level.

Equation (5.1) is a cumulative probability density function approaching 1.0 as the number of agents approaches infinity, that is $\lim_{x \to \infty} \mathcal{S}(x) = 1$. For $\theta \to \infty$, equation (5.1) approximates the unit step:

$$\lim_{\theta \to \infty} \frac{1}{1 + e^{\theta(\tau - x_{\hat{m}})}} \approx \begin{cases} 1 & x_{\hat{m}} > \tau \\ 1/2 & x_{\hat{m}} = \tau \\ 0 & x_{\hat{m}} < \tau \end{cases} \tag{5.2}$$

Although, unlike the unit step function, the limit on the left in (5.2) is always continuous, even at $x_{\hat{m}} = \tau$ where the value of the sigmoid is $1/2$. The proposed model is therefore a generalization of the "stick-pulling" TA model with deterministic team size [20], allowing us to tune the variable resulting group sizes using the tuning parameters τ and θ in (5.1).

Assuming the agents to be loosely synchronized, e.g., by considering decisions within a finite window of time, determining a majority vote corresponds to a Bernoulli trial with each agent flipping a biased coin—the bias being computed using the sigmoid function—to decide whether or not to collaborate in the next time step. The probability that exactly k agents collaborate from a population of n agents at a collaboration site is given by the probability mass function (PMF) of a Binomial distribution.

$$B(n, k) = \binom{n}{k} \mathcal{S}(n)^k (1 - \mathcal{S}(n))^{n-k} \tag{5.3}$$

Since I care about the case when half or more of the agents $(n/2)$ decide to collaborate, the probability $P(n)$ that half or more agents in a group of n collaborate is the cumulative probability of the above PMF from $k = n/2$ to $k = n$.

$$P(n) = \sum_{i=n/2}^{n} \binom{n}{i} \mathcal{S}(n)^i (1 - \mathcal{S}(n))^{n-i} \tag{5.4}$$

This equation describes the probability with which a group of size n at a given collaboration site will decide to successfully collaborate. Note that (5.4) is only an approximation for odd n, which requires rounding $\lceil n/2 \rceil$ to the next integer.

For large group sizes, the Binomial distribution approximates the Normal distribution and (5.4) reduces to

$$P(n) = \int_{n/2}^{n} \mathcal{N}(n\mathcal{S}(n), n\mathcal{S}(n)(1 - \mathcal{S}(n))) \qquad (5.5)$$

Therefore, in a group of size n, and n reasonably high (see below), an average of $n\mathcal{S}(n)$ robots will collaborate with group sizes of variance $n\mathcal{S}(n)(1 - \mathcal{S}(n))$. In the special case of $n = \tau$, i.e., the group size has the desired value of τ, (5.5) evaluates to $P(\tau) = \mathcal{S}(\tau) = 0.5$. Therefore, the probability of a group of n agents to collaborate is identical to the probability of a individual agent to collaborate. In all other cases (5.5) allows us to calculate the micro-macro matching from $\mathcal{S}(n)$ to $P(n)$.

A caveat of (5.5) is that the Normal approximation yields poor results for small n, usually smaller than 20, and is better when $\mathcal{S}(x)$ is neither close to 0 or 1 [65]. In these cases, exact solutions for $P(n)$ require numerical solutions of (5.4) using what is known as *continuity correction* [66].

5.2 Microscopic Model

As the proposed collaboration mechanism are strongly non-linear, I chose microscopic stochastic simulations to explore the underlying dynamics of the system. The approach followed to build the stochastic Gillespie simulation of the system is as follows.

- Perform random walk till a collaboration site is found (*search* state).

- Perform algorithm, Task_Allocation (see Algorithm 1) (*wait* state).

- Complete collective task and disperse. (*collaborate* state).

- Return to search.

The probabilistic finite state machine that describes individual agent behavior for this swarm system is shown in Fig. 5.2. From the individual agent's perspective only one state each exists for *wait* and *collaborate*. From a probabilistic modeling perspective, the wait and collaborate states are meta states, divided into m states each, one for each collaboration site in the arena. This is done to clarify that the probability of collaborating at a given site *only* depends on the number of agents at that specific site and collaborations *only* happen between agents at the same site.

The probability p_{SW_i} in the PFSM model of the system shown in Fig. 5.2 is the probability that an agent encounters a collaboration site. This is geometrically computed as the ratio between the total area of the search space (arena) and the total area of collaboration sites, i.e. $p_{SW_i} = n_s(A_s)/A$ (n_s = number of sites, A_s = area per site). The probability, $P_{W_iC_i}$, of going from a wait state to a collaboration state is given by eq. (5.4) with input N_{W_i}, the number of agents at collaboration site-i. P_{C_iS} stochastically models the time it takes for an agent to complete a generic collaborative task and is equal to $1/T$, where T is the amount of time (on average) that it takes an agent to complete the collective task. Note that agents have a zero probability of transitioning from the *wait* state back to the *search* without collaborating, i.e. once an agent is at a collaboration site, it will not leave till a collaboration event happens at that site. I chose the following numerical values for all simulations, unless otherwise noted: $A = 100cm^2$ and $A_s = 10cm^2$.

For the sake of simplicity, consensus between agents—i.e. going from W_i to C_i—at the same collaboration site is assumed to happen instantly and therefore the extra state(s) is/are omitted from robot controller.

In order to compare the dynamics of the proposed probabilistic TA mechanism with the deterministic one by Lerman et. al [20], I implemented a variation of the above algorithm using a unit-step at τ instead of the sigmoid function and removing the consensus step, which is not necessary in this model.

I use Gillespie simulation [41] to explore the dynamics of the proposed collaboration model. For both experiments a single collaboration site is used and each run simulates 300s of time. The

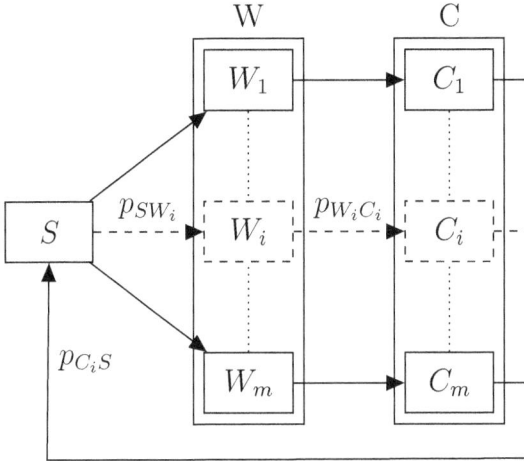

Figure 5.2: Agent controller used to drive group collaboration. There is a Search state and m Wait and Collaboration states, W_i and C_i respectively— one for each collaboration site.

desired group size (τ, in Eq. (5.1)) is set to 4, 8, 16 and 32 agents out of a total of 100 robots. The collaboration task is programmed to take 10s, on average, per agent. Data points are gathered by averaging data from 100 identically set up runs in each case. The *rate of collaboration* for the threshold model is computed by summing the number of groups that successfully collaborate and dividing by the total experiment time (300s). For the deterministic model, collaboration rate is computed by summing all successful collaborations, i.e. collaborations involving team sizes equal to τ, and dividing the the experiment time (300s).

5.3 Experiments and Results

I will first compare the dynamics of the proposed approach with Lerman et al.'s k-collaboration model [20] and then validate the emergence of group sizes with similar means but varying variances. Figure 5.3a shows collaboration rates for both models when θ is set to 2 (for the probabilistic model) and the wait time is set to ∞ (for the deterministic model), in order to allow for a fair comparison. (All experiments are run in a regime where infinite wait times are

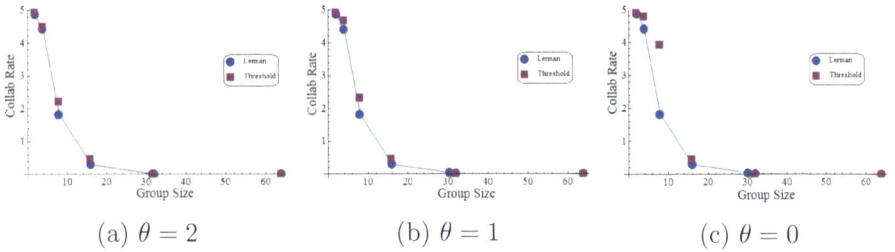

Figure 5.3: Comparison of the collaboration rate for TA with probabilistic and deterministic [20] for different values of θ and team sizes τ in an environment with one collaboration site and one hundred robots.

optimal wait times, i.e., there are more agents than collaboration sites.) Figures 5.3a, 5.3b and 5.3c show collaboration rates for $\theta = 2$, $\theta = 1$ and $\theta = 0$ with infinite wait time. With $\theta = 0$, the logistic function is uniformly 0.5, allowing any team size to form. With increasing θ the logistic function approximates a unit step, minimizing the variance.

I observe the collaboration rate to be qualitatively and quantitatively very similar for high values of θ (steep slope), and to exceed that of the deterministic model for very low values of θ (flat slope). This is expected as flat slopes increase the variance of the observed group size and therefore allow much smaller teams than τ agents to collaborate.

Figure 5.4 shows histograms of the resulting group sizes for various values of $\tau = 4, 8, 16, 32$ and $\theta = [0; 0.1; 1]$ (100 simulations per data point). It is clearly seen that when θ is set to 0, the sigmoid becomes constant $(\mathcal{S}(x) = 1/(1 + e^0) = 0.5)$ so agents have an equal probability to want to collaborate or not, no matter what the desired group size is. I therefore see a large number of small groups forming, with most groups consisting of 2 agents. This is to be expected since the expected number of agents willing to collaborate in a group of size 2 is 1, given the probability of collaboration is constant at 0.5.

Figure 5.5a displays average group sizes as θ is varied from 0 to 1 and τ from 4 to 32 based on the data from Figure 5.4. I observe that for large enough values of θ the mean of the group size distribution approaches the desired group size and is largely unaffected by increasing θ. Thereafter, its magnitude depends only

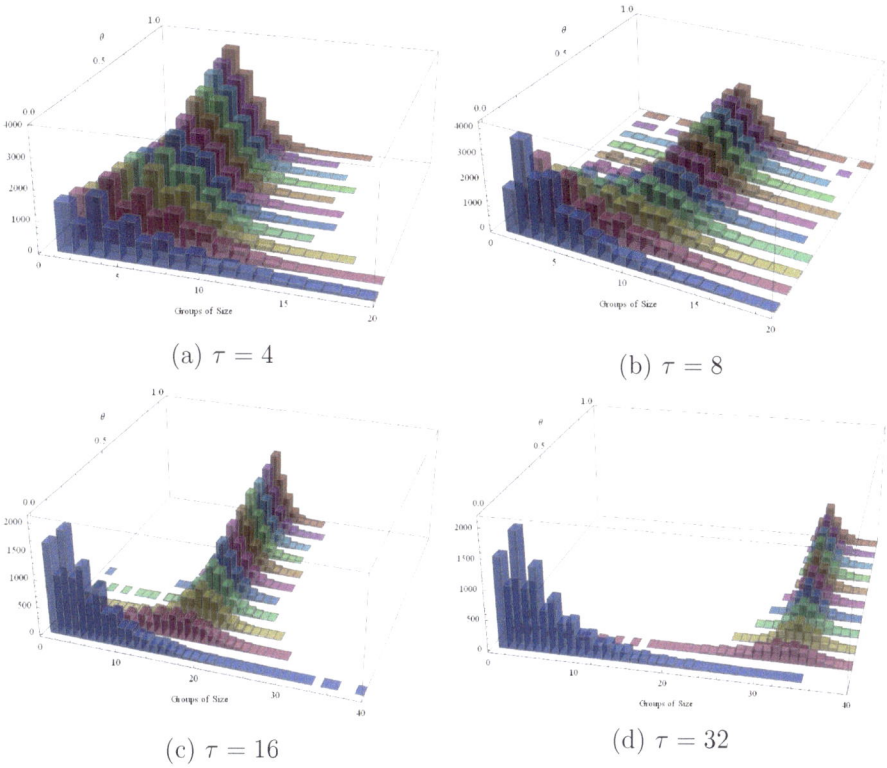

(a) $\tau = 4$

(b) $\tau = 8$

(c) $\tau = 16$

(d) $\tau = 32$

Figure 5.4: Histograms of resulting team sizes for various values of τ and θ with one hundred robots and one collaboration site.

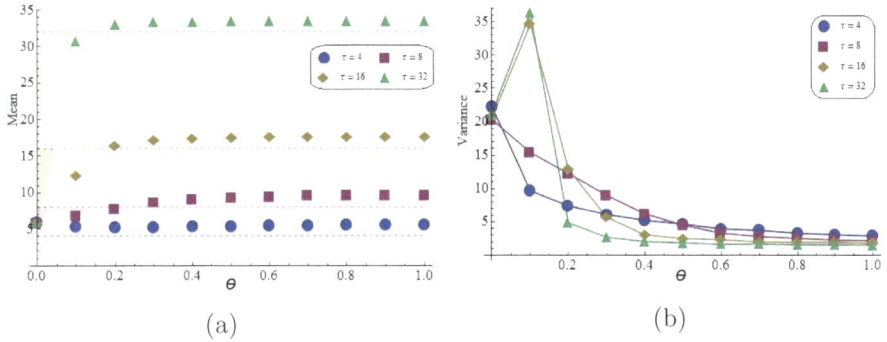

(a)

(b)

Figure 5.5: Showing the effects of varying θ on means and variances corresponding to the histograms seen in Fig. 5.4.

on τ except in the special case where $\theta = 0$ where it is constant. The relative error of the mean compared to the desired average decreases with increasing number of agents as the Binomial distribution (5.4) approximates the Normal distribution (5.5).

Figure 5.5b shows how the variance of group size decreases with increasing θ. This is because the sigmoid function approximates the unit step, making the team size more and more deterministic. On the other hand, low values of θ lead to large variances in the group size. For $\theta = 0$, the variance is constant for all values of τ and depends exclusively on the total number of robots.

Figure 5.6: (Left) The Droplets platform set up for the response threshold model experiment to measure group sizes by sweeping τ and θ. (Right) A single droplet robot without it's top cover showing omnidirectional IR sensors for communication.

Finally, I use the Droplet swarm robot platform to perform real experiments to study the effects of using the proposed TA scheme on a physical system. The Droplets are small individually simple robots capable of omni-directional motion and communication (via IR) as well as sensing patterns projected from above as seen in Fig. 5.6. In my experiment I assume that all agents have already arrived at a collaboration site and measure the corresponding collaboration rates for a team of 6 robots while varying values of τ and θ. Each agent is individually running the algorithm described in Alg.1. A collaboration event is recognized by having all the robots turn on their green LEDs for 5 seconds. After such a collaboration event, each agent resets its group size estimate and runs Alg.1 again.

I ran 5 repeated experiments for all 15 combinations of $\tau =$

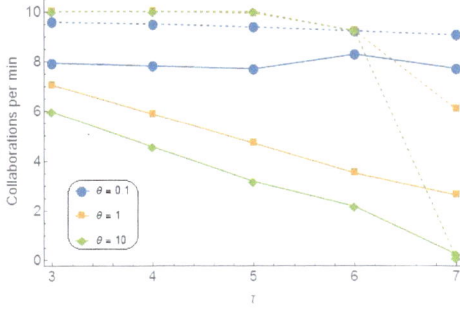

Figure 5.7: The solid lines show collaborations per min, over 15min, for a group of 6 real robots as the desired group size is varied from 3 to 7 and the slope of $\mathcal{S}$ is varied between $0.1, 1$ and 10. The dashed lines indicate simulation results with the same parameters.

$3, 4, 5, 6$ and 7, and $\theta = 0.1, 1$ and 10, totally 75 runs. Each experiment lasted 15 minutes and an overhead camera system was set up to detect collaboration events using the software *RoboRealm*. The collaboration rate was a value computed by counting the number of collaborations over the course of each 15 minute experiment, normalizing to collaborations per minute, and averaging over the 5 repeated runs. To account for the vision software's detection errors, the raw data gathered from each experiment was de-bounced and passed through a low-pass filter to expose real collaboration events while eliminating observation error. The results of these experiments are seen in Figure 5.7. While results are in accordance with simulation for θ being low, the collaboration rate on the real robot platform is much lower than expected for larger θ as simulation assumes perfect communication and group size estimates.

5.4 Discussion

Results in Figures 5.3, 5.4 and 5.5 show that the proposed threshold-based TA mechanism is a generalization of the deterministic Lerman model in that it allows to approach what is seen with deterministic group sizes while retaining the elasticity to vary group sizes along any desired range of values. Also, these plots show how altering microscopic control parameters within the agents, θ

and τ of their sigmoid functions, directly affects macroscopic behavior of the swarm system by altering means and variances of formed group sizes, respectively. Although the matching between microscopic results and macroscopic prediction is not perfect due to the discrete approximation, the plots show that a wide range of means and variances are feasible. Finding appropriate parameters to reach these could be easily achieved using a suitable optimization framework such as presented in [22,23], using the macroscopic predictions as initial estimate.

The proposed TA algorithm requires an estimate of the group size at each collaboration site as well as the ability to communicate with the group in order to reach a consensus. While these assumptions seem to be limiting at first sight, they can be rolled into the analysis process and possibly exploited to design the TA process. For example, an increasing variance for observing the group size τ or noise in the consensus process simply increase the variance of the TA process and could therefore be countered — to some extent — by altering the properties of the CRT function.

This effect is clearly observed in the physical experiment results (see Figure 5.7). Since the communication between real robots is not perfect, they almost always underestimate the size of their group resulting in lower collaborations for high θ and τ values. As I observe from comparing the micro simulation results—that are modeled with perfect communication—with real experiment data, I observe a large discrepancy when $\theta = 10$. This happens because although individual agents are set up to be in a group of size 6, their estimates for the group size never cross 4 due to imperfect and blocked communication. Coupled with the fact that the sigmoid threshold effectively acts as a step function when $\theta = 10$, this results in approx. 0 probability of collaboration between agents for a desired group size of 6 but a group size estimate of ≤ 4. Lower values of θ result in better matching between real and simulation data since lower slopes effectively increase the variance in allowed group sizes and mitigate this effect.

I note that there is no optimal wait time as in stick pulling-like collaboration [20]. This optimum exists in swarms with less robots than sticks, which is shown analytically in [21]. Such an optimum does not exist in the proposed model as there is a non-zero probability team sizes with $n < \tau$ will eventually collaborate. Indeed, Algorithm 1 eventually completes as $\mathcal{S}(x) > 0 \forall x$, i.e., even if only very few robots are at a collaboration site and τ is large, there is a non-zero probability that half or more of the agents at the site eventually collaborate (see also Equation 5.4).

Although the algorithm does not deadlock—the probability to collaborate even if the team size is far off the desired value— the resulting behavior might be undesirable, resulting in potentially very long wait times and poor task performance. This could be mitigated by introducing preferential detachment from small groups and preferential attachment to larger groups as customary in swarm robotic aggregation [67].

In practice, effective collaboration rates will also be limited by the embodiment of the robots, which might make finding physical space at a site cumbersome. In the presented microscopic simulation, for both stochastic and deterministic team sizes, the number of robots per site were not limited, allowing scenarios in which multiple groups collaborate in quick succession at the same site. While comparing both models without embodiment is reasonable, I wish to study the effect of embodiment in future work.

5.5 Summary

In this chapter I presented a multi-agent collaboration algorithm to recruit an approximate number of individually simple robots with controllable variance. I proposed a sigmoid RT function motivated by TA in social insects and describe macro-level models backed by micro-level simulations to predict the resulting team sizes and their variance. These results were further validated through physical experiments using the *Droplet* swarm robotics platform. I showed that the slope of the CRT function could be used to control the variance of group size, allowing agents to trade off deterministic team size with coordination speed, and making the proposed mechanism applicable to a variety of applications.

Chapter 6

Comparing Centralized vs. Hybrid Approaches to Task Allocation

Chapter 3 sets up a formal model for defining task allocation in multi-agent systems. In doing so, it also presents a valid definition of optimal task allocation by formulating the problem as an ILP. The solution to this ILP is a mapping from agents to targets that result in maximized system utility. Therefore, a natural next step would be to design a controller for the system that could communicate with agents and provide them with these optimal assignments.

For such a centralized approach to be viable, attention must once again be drawn to the fact that the definition of optimality assumes perfect information about the state of the system. All object positions as well as target magnitudes τ_t must be known the to the central controller to construct an accurate objective function and constraints. If a central controller has incomplete or inaccurate information about a system then it is very likely the assignments returned by the ILP could result in significant wastage of resources by agents. In such a situation we see how the RT algorithm comes into play to prevent unnecessary task attempts by a group of agents. This chapter focuses on comparing such a centralized approach with a hybrid approach which uses the RT strategy from the previous chapter layered upon noisy centralized control.

6.1 Centralized Optimal Allocations

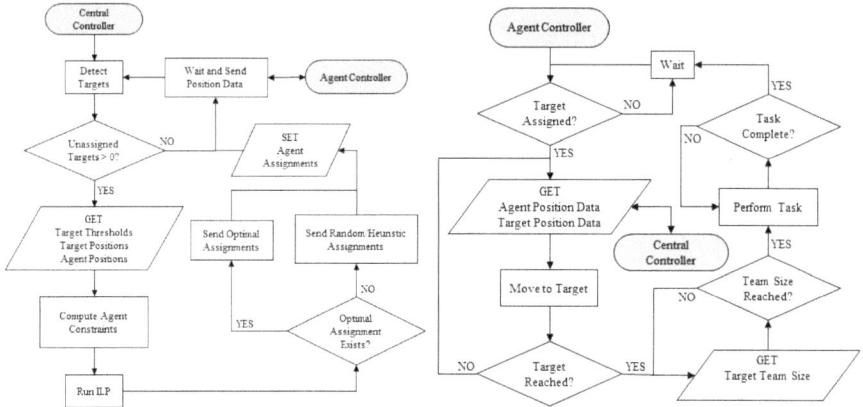

Figure 6.1: Left flowchart describes the control flow for the central controller of the centralized optimal assignment method. Right flowchart describes the control flow for the individual agent controller of the centralized optimal assignment method.

Using the problem formulation from Chapter 3 we can pose the TA optimization problem as a 0-1 ILP. This means that finding an optimal assignment of agents to targets (taking constraints into account) can be achieved by using any of the many existing methods for solving ILPs. This also means that the TA optimization problem exists in the realm of NP-hard problems and no polynomial time algorithm currently exists for finding such an optimal assignment. Although, efficient methods do exist for finding near optimal solutions. Also, given a constrained domain such as a set number of agents and relatively few targets most existing computers can find TA solutions well within the physical movement time-frame of robots.

The caveat for using a centralized algorithm is that the controller is assumed to have perfect knowledge of the operational environment and model parameters for the tasks being accomplished. Since this information is almost always subject to the accuracy constraints of the sensors making on-field measurements

a centralized method is not always feasible. Nonetheless, we study
the effects of using a centralized controller assuming perfect system
information as a baseline — being the best possible performance
a multi-agent system can achieve for TA.

Fig. 6.1 describes how the centralized controller works and
communicates with each agent's on-board controller. Target and
agent information is fed to the controller to set up the ILP. The
values of importance here are task magnitudes τ_j, target specific
welfare $W(t_j) = w_j$ if $|a_j| \geq k_j$ and 0 otherwise, and target thresh-
olds $K(t_j)$ or k_j. We set,

$$w_j = \tau_j - \sum_{n \in a_j} dist(p, t_j)$$

where $dist(n, t_j)$ is the normalized straight-line distance between
agent n ($\in a_j$) and target t_j. To recall, a_j is the set of all agents assigned to target t_j by an assignment. Taking distance into account
mimics agent resources in a way, as the controller now exhibits
spatial awareness of the system and can account for it when pre-
scribing allocations. Target thresholds are assumed to be equal to
task magnitudes for the purposes of this experiment, i.e., $\tau_j = k_j$,
$\forall t_j \in \mathcal{T}$.

Recalling the ILP constraints and objective to maximize are:

$$\text{Maximize} \quad \mathcal{W} = \sum_{t \in \mathcal{T}} W(t, |\mathcal{A}(t)|) \quad\quad (6.1)$$

$$\text{S. T. for all} \quad i = 1 \ldots |\mathcal{P}|, j = 1 \ldots |\mathcal{T}|,$$

$$\sum_{t \in \mathcal{T}} \xi(n_i, t) \leq 1$$

$$\sum_{t \in \mathcal{T} \backslash C(n_i)} \xi(n_i, t) = 0$$

$$0 \leq \xi(n_i, t_j) \leq 1$$

The first condition ensures that all agents get assigned to at most
one target. The second condition accounts for each agent's target
constraints c_i and essentially locks in the value of $x_{ij} = 0$ for all
$t_j \notin c_i$. The final constraint makes this optimization problem a
0-1 ILP since all x_{ij} can only have the value 0 or 1, indicating
whether or not agent i is assigned to target j.

6.2 Distributed Response Threshold based Task Allocation

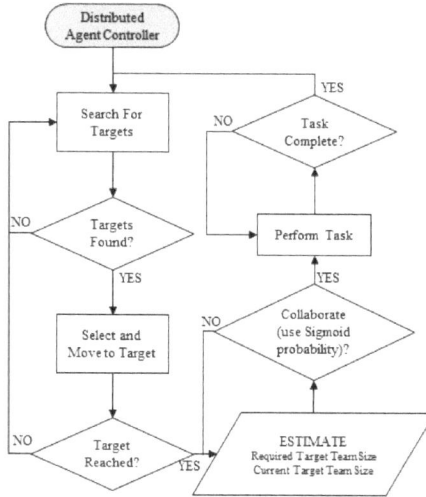

Figure 6.2: This flowchart describes the control flow for an indiviaul agent's controller in the de-centralized RT based optimal assignment method.

When the central controller stops receiving perfect information about the system it's behavior quickly becomes non-ideal. For example, if the perception of task magnitude τ_j becomes noisy then agents are often assigned non-optimally and resources are wasted. The RT TA strategy can be employed to mitigate this effect. Once again taking the firefighting scenario into account. If a non-ideal team size is assigned to a task/fire by a central controller due to imperfect estimation of fire size, then it should still be viable to hold off from starting the containment process until a more reasonable number of agents arrives at the fire. This is important as agents have only finite resources to complete their assigned tasks. In the case of firefighting, this resource may be the amount of fire repellent being carried on energy stores of the robot.

The flowchart in Fig. 6.2 shows a contrasting agent based controller when compared to the one shown in Fig. 6.1. In both cases, selection and movement information is provided by a central entity but in the case of Fig. 6.2 agent do not immediately initiate the task they are assigned to. Instead, they perform the RT strategy discussed Chapter 5. Each agent independently determines, based on their own measurements of task magnitude, whether or not to concurrently begin the task. The same majority voting algorithm presented earlier is employed to achieve concurrency. The RT function used is the continuous logistic function from Eq.5.1. τ from Eq. 5.1 is set based on each individual agent's noisy perception of the task's true magnitude τ_j, i.e. $\tau = \mathcal{N}(0,1) + \tau_j$. We assume standard Gaussian noise on agent sensors for these experiments but any desired noise value could be used here, such as one measured when calibrating a real robot's on-board sensors. The input to the logistic sigmoid function is the number of other agents at that particular task site, $|a_j|$ which is also made available to an agent via indirect (stigmergic) or direct communication, such as during the voting process or by the central controller. The performance of this hybrid approach compared to the completely centralized approach from the previous section is detailed and analyzed in the following sections of this chapter.

6.3 Experiments

Designing experimentally verifiable and reproducible metrics for comparing pure distributed TA with an optimal centralized controller is challenging. This is because optimal centralized control really performs two functions, it assigns particular agents to particular targets and in doing so it also optimizes the utility for the entire swarm system. Distributed TA, on the other hand, assumes that agents are already at task sites and answers the question of whether or not an individual should perform a task based so as to also optimize macro-level utility and minimize wastage of resources.

The experiment described herein attempts to enforce a fair comparison between both approaches. Reusing to the firefighting scenario from [12], the field is set up with five "fires" or target sites, as seen in Fig. 6.3. Each target has a varying magnitude-τ_t, indicating the number of agents required to successfully complete it. The magnitudes are chosen such that $\sum_{t \in [1,...,5]} \tau_t = N$ where

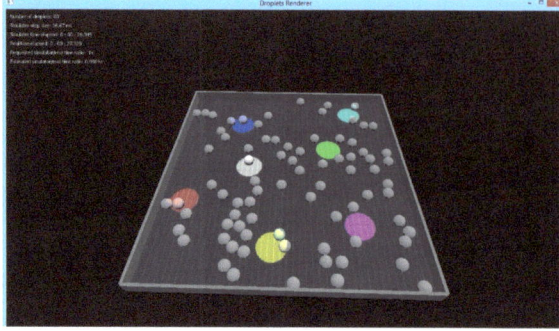

Figure 6.3: Simulated experiment setup. The colored circles are task locations with different magnitudes. All robots are assigned to tasks based on centralized on distributed control strategies, depending on the experiment being run.

N is the total number of agents. For this experiment N was set constant to 20 and because of this condition uniformity in randomness for picking τ_t between $[2, 20]$ is not guaranteed. Having the exact number of agents required to successfully complete all tasks provides a normalized starting condition for comparing all the controllers. This setup can be thought of as a "snapshot" of a realistic situation where targets' magnitudes change with time and task allocation needs to be redone at regular time intervals. The temporal aspect of real-world situations is not considered in these experiments as it strays for the objective of comparing the effectiveness of central vs. distributed controllers under the same conditions.

Three different controllers are considered. The first controller is a centralized ILP solver that is assumed to have perfect knowledge of the system, i.e. it knows exactly where agent and fires are located in the experimental arena and also has perfect knowledge of every task's magnitude. This controller is used as a baseline optimum to compare the other controllers to. We can reasonably expect any TA strategy to do no better than this optimal controller. The next controller is also a centralized ILP solver with perfect information about agent and target locations but measures task magnitude with Gaussian noise $\mathcal{N}(0, \sigma)$ added. σ is set to $1, 2, and 5$, dubbed low, medium, and high noise, respectively, for three distinct sets of experiments to observe the difference in performance between the central vs. hybrid controllers. At the start of each experiment all 20 agents and 5 targets are assigned uniformly random locations. Each controller is run on the same initial conditions.[1] Finally, a hybrid controller is considered such that the actual agent assignments still come from the noisy central controller but once agents are at a particular they run their own RT algorithm to decide whether or not to attempt the task. The expected advantage here is that clearly bad assignments may be mitigated. Agents' on-board sensors are also assumed to be noisy with Gaussian noise $\mathcal{N}(0,1)$ added to their independent estimates of the task magnitude (and consequently desired team size), τ_t. We set the slope parameter θ of the logistic function used by all agents to 1 for the sake of simplicity in analyzing the results for these experiments. Tuning θ based on calibrated sensor and communication error is a very interesting avenue to explore for future work in improving RT based TA but is not shown here. The results of running this series of experiments on all three controllers are presented below.

All experiments were carried out on a 64-bit PC running on an 8-core AMD processor with 16GB of high-speed DDR3 memory. The ILP solver used was compiled on 32-bit architecture and implemented via a 32-bit python (2.7) API so the memory restriction per process were fixed at 4GB. Interestingly, 4GB of memory was often not enough for the ILP solver to find a solution to the optimization problem from Eq. 3.4 even for only 20 agents and 5 targets. Simulations runs had to sometimes be restarted and a single successful run would often take 1–5 minutes even on a high-end machine. While optimizations to the code may have alleviated some of these problems it was interesting to observe a high-end machine having difficulty with the relatively small system being studied here. A centralized controller for a realistic scenario involving hundreds or even thousands of agents and tens or hundreds of tasks would likely required considerable computing resources to solve in a timely manner.

---

[1]The code used to run these simulation experiments is available at: www.github.com/akanakia/collab-optimizer. The optimizer used is called *z3*. It is an open source (MIT license) SMT solver developed at Microsoft Research (https://github.com/Z3Prover/z3).

6.4   Results

One hundred experiment sets were run on each controller. Each experiment had different initial conditions but the conditions were consistent between all three controllers. The results presented in Fig. 6.4 compare the average number of tasks attempted per run to avg. failed task assignments per run for each controller. A task was considered assigned if at least one agent was assigned to it. A task was considered failed if at least one agent was assigned to it but fewer than k_j — the true target threshold — and the agents attempted that task. For the central controllers cases assigned agents automatically attempted the task whereas for the distributed case, agents first ran the RT algorithm to decide whether or not to attempt a task that they were assigned to. This way of representing the data allows us to easily compare the rates of resources wasted per experiment and per agent as well as the average number of tasks successfully assigned by a controller and attempted by agents.

We see some interesting behavior when applying the RT strategy to agents after they have been assigned tasks. First off, the central controller with perfect (ideal) system information has an avg. assignment rate of *nearly* 100% for agents and targets. It is interesting to note that this value sometimes dropped below 100% because in some rare cases, the optimal ILP solution involved not assigning any agents to a target even though agents were available. There were also some cases where the ILP solver terminated before any solutions could be found.

Looking at the central noisy controller, the percentage of failed tasks was right around 35% in all cases from low noise addition, $\sigma = 1$ to more noisy values when $\sigma = 5$. Although, the percentage of agents failed did lower from about 35% when $\sigma = 1$ to about 19% when $\sigma = 5$. With the notable exception of total tasks attempted in a high noise environment, the noisy central controller's tasks/agents attempted were comparable to the central controller in all cases. This resulted in a high attempt rate but high failure mode for the centralized noisy controller.

One the other hand, the distributed controller fared much better in the number of failed task attempts, never attaining failure

Tasks	Attempted	Failed	Attempted %	Failed %
Cen. Ideal	4.73	0.16	94.6	3.38266
Cen. Noisy	4.76	1.74	95.2	36.5546
Distr. Noisy	2.52	0.34	50.4	13.4921

Target Data (Central Noise $\sigma=1$)

Agents	Attempted	Failed	Attempted %	Failed %
Cen. Ideal	19.12	0.24	95.6	1.25523
Cen. Noisy	18.01	6.28	90.05	34.8695
Distr. Noisy	10.03	1.15	50.15	11.4656

Agent Data (Central Noise $\sigma=1$)

Tasks	Attempted	Failed	Attempted %	Failed %
Cen. Ideal	4.74	0.17	94.8	3.5865
Cen. Noisy	4.36	1.58	87.2	36.2385
Distr. Noisy	2.51	0.38	50.2	15.1394

Target Data (Central Noise $\sigma=2$)

Agents	Attempted	Failed	Attempted %	Failed %
Cen. Ideal	19.03	0.21	95.15	1.10352
Cen. Noisy	18.2	4.83	91.	26.5385
Distr. Noisy	12.24	0.99	61.2	8.08824

Agent Data (Central Noise $\sigma=2$)

Tasks	Attempted	Failed	Attempted %	Failed %
Cen. Ideal	4.84	0.1	96.8	2.06612
Cen. Noisy	3.96	1.36	79.2	34.3434
Distr. Noisy	2.39	0.22	47.8	9.20502

Target Data (Central Noise $\sigma=5$)

Agents	Attempted	Failed	Attempted %	Failed %
Cen. Ideal	19.47	0.11	97.35	0.564972
Cen. Noisy	18.24	3.55	91.2	19.4627
Distr. Noisy	14.17	0.63	70.85	4.44601

Agent Data (Central Noise $\sigma=5$)

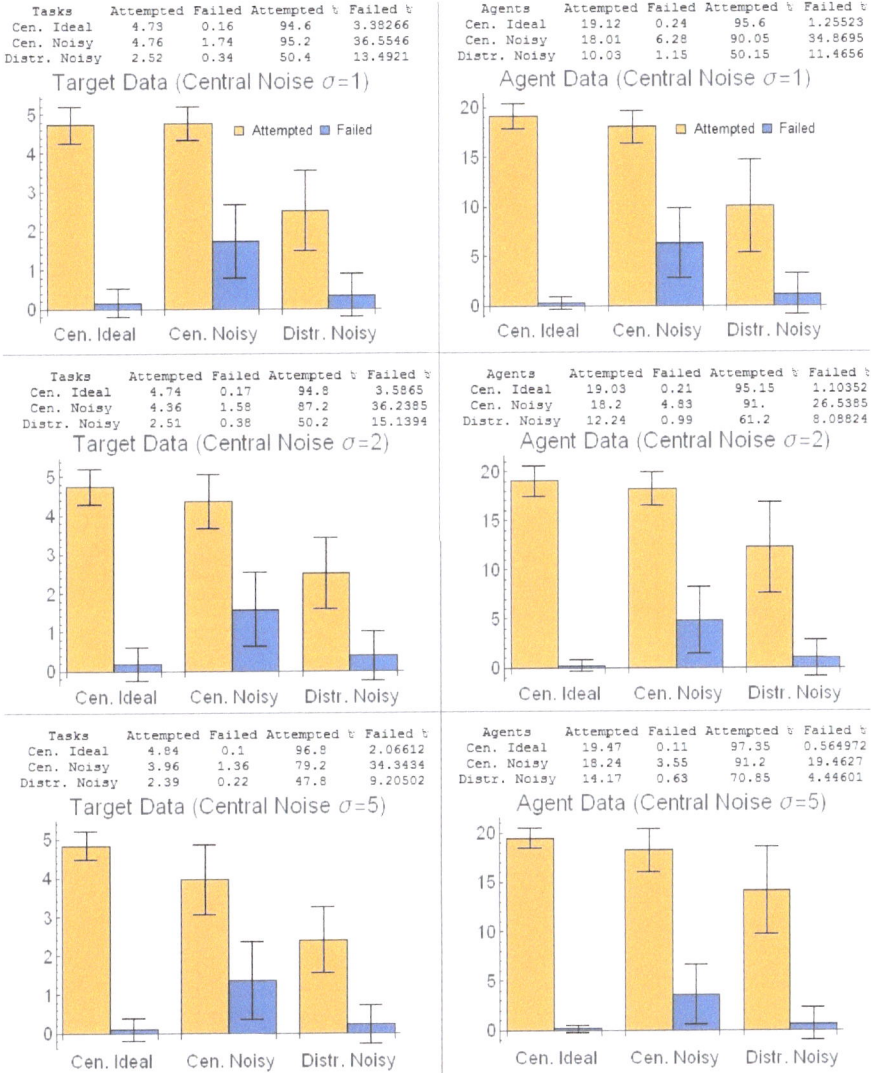

Figure 6.4: Simulation experiment results for three types of control — central ideal, central noisy, and hybrid noisy. Yellow bars in the top row show avg. number of attempted tasks with std. deviation error bars. Blue bars show the avg. number of attempted tasks that failed. Bottom row show similar results from an agent-centric viewpoint, i.e. assigned vs. failed agent averages. 100 experiment runs per controller for each bar.

rate higher than 16% for tasks or agents. This lower failure rate did come at the cost of lower total task attempts but the ratio of attempted to failed tasks or agents was consistently higher for the distributed controller than the noisy centralized controller, under the same assignment conditions.

6.5 Discussion

The first result to discuss is the below 100% success rate of the central ideal controller. Two major factors attributed to this result. Firstly, the fact that spatiality was considered by the solver and made it sometimes unfeasible to move agents to their closest targets since they may have better served larger targets that were farther away. Therefore, in some rare cases targets may have been left completely unassigned. Perhaps, more surprising was the fact that some assignments resulted in failed attempts meaning the ideal controller was not able to assign enough agents to a task. The reason for this is that the ILP solver halted at a sub-optimal solution and returned that as a viable solution once the recursion limit was hit while exploring the search space for better solutions. This happened in extremely rare circumstances and may have been avoided by increasing the memory limit of the solver or via other code optimizations.

The centralized noisy controller exhibited relatively high attempt rates for both tasks and agents and was comparable to the ideal case. This comes at the expense of a relatively high failure rate. The consistently high attempt rate may also be a statistical side effect of a relatively low number of agents and, therefore, agent distributions. The additive noise term was applied to each task magnitude value but these values are still floored at two agents since it does not make sense to have collaborative task assignments with fewer than two agents. This means that task magnitude and consequently group sizes are unevenly skewed towards higher values since negative error for small group sizes is capped at two. Nevertheless, these were the same distributions provided in the distributed hybrid controller case and it fared considerably better. Even though fewer tasks were attempted overall by the agents, we can see a noticeable trend towards increased attempt rates and lower failure rates even as overall noise increases in the system. In the case of high central noise, we see an agent attempt rate of around 70% and a failure rate of only 4% making it close to optimal. This, compared to using just the central noisy approach which results in about 90% agent attempt rate but a much higher 19% failure rate. These results show how a hybrid approach relying on some (very) noisy information from a central controller can be substantially improved by applying the RT distributed algorithm for TA.

It is worth reiterating at this point that both, the agents' additive noise term ($\mathcal{N}(0, 1)$) as well as the slope of the sigmoid function ($\theta = 1$) were kept consistent throughout the course of this experiment. A very interesting avenue for future research would involve designing an strategy of iterative refinement of θ to compensate for agent sensor noise. Such a approach is discussed in [68]. Simulation experiments such as the ones presented here could form the basis for the iterative refinement model employing real robots and tuning their control parameters in the field.

Chapter 6. Centralized vs. Hybrid Approaches

Chapter 7

Conclusion

Task allocation is a well studied problem in the fields of robotics, optimization, and game theory where a number of identical agents must be assigned to the a number of collaborative tasks for maximum system welfare. Swarm robotics in particular has tackled this problem for a number of years and many centralized, distributed, and hybrid approaches exist for solving task allocation. Practical deployment of multi-agent systems for automated surveillance, robotic firefighting, and oil-spill containment among other tasks have been proposed as a viable alternative to existing approaches and corresponding task allocation strategies for each of these applications have been analyzed but there has so far been no formal unifying definition for optimal task allocation in swarm robotics. This dissertation therefore presented a formal problem definition for multi-agent task allocation as well as a general definition of optimal task allocation as an integer linear program when all information is known about the system.

The center of focus of this dissertation is a response threshold based distributed task allocation strategy. Response threshold strategies have been proposed for multi-agent task allocation since ethology research [25, 26, 49] provides a phenomenological argument for their simplicity and reliability in social insects. So far as I am aware, a mathematical argument for their widespread prevalence in the fields of ethology and swarm robotics has been lacking. To fill in this gap, I along with Dr. Behrouz Touri present a game theory inspired argument for their use. The widely applicable concept of concurrent benefit tasks is also defined here. This work suggests that global games can describe a wide range of collective decision making scenarios, explaining the prevalence of sigmoid threshold strategies in natural and artificial systems. By varying noise in the system and thereby changing the slope of the sigmoid function, the response threshold strategy also allows the system to balance between exploitation and exploration. This, in turn, may lead to designing robust robot swarms that are flexible and can alter strategies for changing environmental parameters without requiring communication; a highly desired property that is often observed in natural swarms and is of great interest for engineered systems.

This work is followed by the actual response threshold based task allocation algorithm, itself. The algorithm allows to recruit an approximate number of agents with a desired variance to a task requiring team sizes centered around some mean magnitude value, using a logistic sigmoid function. This allows to trade-off task execution accuracy with speed, resulting in increasing collaboration rates for teams with larger variances. I demonstrate that task allocation of teams with deterministic size is a subset of the proposed stochastic task allocation mechanism. As such, the proposed framework provides a computationally simple, adaptive, and robust alternative for centralized coordination.

A centralized approach is then compared as a baseline optimum to a proposed hybrid task allocation method. While the distributed strategy for task allocation presented earlier provides a robust means of making decisions on concurrent benefit tasks, it does not plan for agents to move to tasks. The centralized optimal controller discussed earlier achieves this by taking agent and task positions into account while solving the ILP. While a centralized approach to task allocation is always optimal when provided with perfect information about the real-world system, in realistic scenarios involving agent and task constraints and imperfect sensing and communication we see that the distributed approach quickly becomes extremely useful to minimize failed task attempts and reduce wastage of resources (or maximize utility, as it is defined in the field of optimization). The results of these experiments show that when a noisy central controller is used alongside the response threshold based task allocation strategy, we see fewer attempted tasks but also a disproportionate reduction in the number of failed task attempts. In fact, the experimental data would suggest that as the central controller gets increasingly noisy data about task magnitudes, employing the a distributed approach results in an increased rate of successful task assignments. This result falls in line with predictions of an Bayesian Nash equilibrium existing when agents receive a common noisy stimulus signal and employ threshold strategies to decide whether or not to participate in the task. The results from the final chapter provide a very compelling reason to further research and extend the results provided in this dissertation.

Bibliography

[1] G. Beni, "From swarm intelligence to swarm robotics," in *Swarm Robotics* (E. Şahin and W. M. Spears, eds.), no. 3342 in Lecture Notes in Computer Science, pp. 1–9, Springer Berlin Heidelberg, Jan. 2005.

[2] E. Şahin, "Swarm robotics: From sources of inspiration to domains of application," in *Swarm Robotics* (E. Şahin and W. M. Spears, eds.), no. 3342 in Lecture Notes in Computer Science, pp. 10–20, Springer Berlin Heidelberg, Jan. 2005.

[3] M. Brambilla, "Group size estimation in swarm robotics," 2013.

[4] L. Bayindir and E. Şahin, "A review of studies in swarm robotics," *Turkish Journal of Electrical Engineering*, vol. 15, no. 2, pp. 115–147, 2007.

[5] A. F. T. Winfield, C. J. Harper, and J. Nembrini, "Towards dependable swarms and a new discipline of swarm engineering," in *Swarm Robotics* (E. Şahin and W. M. Spears, eds.), no. 3342 in Lecture Notes in Computer Science, pp. 126–142, Springer Berlin Heidelberg, Jan. 2005.

[6] A. Martinoli and F. Mondada, "Collective and cooperative group behaviours: Biologically inspired experiments in robotics," in *Proceedings of the Fourth International Symposium on Experimental Robotics*, pp. 3–10, Springer Verlag, 1995.

[7] A. Martinoli, A. J. Ijspeert, and F. Mondada, "Understanding collective aggregation mechanisms: From probabilistic

modelling to experiments with real robots," *Robotics and Autonomous Systems*, vol. 29, no. 1, pp. 51–63, 1999.

[8] W. Agassounon, A. Martinoli, and R. Goodman, "A scalable, distributed algorithm for allocating workers in embedded systems," in *Systems, Man, and Cybernetics, 2001 IEEE International Conference on*, vol. 5, p. 33673373, 2001.

[9] A. J. Ijspeert, A. Martinoli, A. Billard, and L. M. Gambardella, "Collaboration through the exploitation of local interactions in autonomous collective robotics: The stick pulling experiment," *Autonomous Robots*, vol. 11, no. 2, p. 149171, 2001.

[10] W. Agassounon and A. Martinoli, "A macroscopic model of an aggregation experiment using embodied agents in groups of time-varying sizes," in *Proc. of the IEEE Conf. on System, Man and Cybernetics (SMC), Hammamet, Tunisia*, pp. 250–255, 2002.

[11] K. Sugawara, N. Correll, and D. Reishus, "Object transportation by granular convection using swarm robots," in *Distributed and Autonomous Robotic Systems (DARS)*, 2012.

[12] A. Kanakia and N. Correll, "A response threshold sigmoid function model for swarm robot collaboration," in *Distributed and Autonomous Robotic Systems (*Best Paper*, DARS)*, 2014.

[13] N. Correll, *Coordination schemes for distributed boundary coverage with a swarm of miniature robots: synthesis, analysis and experimental validation.* PhD thesis, Ecole Polytechnique Fdrale, Lausanne, Switzerland, 2007.

[14] G. Beni and J. Wang, "Swarm intelligence in cellular robotic systems," in *Robots and Biological Systems: Towards a New Bionics?*, pp. 703–712, Springer, 1993.

[15] B. P. Gerkey and M. J. Matarić, "A formal analysis and taxonomy of task allocation in multi-robot systems," *The International Journal of Robotics Research*, vol. 23, no. 9, pp. 939–954, 2004.

[16] P. Amstutz, N. Correll, and A. Martinoli, "Distributed boundary coverage with a team of networked miniature robots using a robust market-based algorithm," *Annals of Mathematics and Artificial Intelligence*, vol. 52, no. 2-4, pp. 307–333, 2008.

[17] L. Vig and J. A. Adams, "Coalition formation: From software agents to robots," *Journal of Intelligent and Robotic Systems*, vol. 50, no. 1, pp. 85–118, 2007.

[18] J. Chen and D. Sun, "Resource constrained multirobot task allocation based on leader–follower coalition methodology," *The International Journal of Robotics Research*, vol. 30, no. 12, pp. 1423–1434, 2011.

[19] K. Dantu, S. Berman, B. Kate, and R. Nagpal, "A comparison of deterministic and stochastic approaches for allocating spatially dependent tasks in micro-aerial vehicle collectives," in *IEEE/RSJ International Conference on Intelligent Robots and Systems (IROS)*, pp. 793–800, 2012.

[20] K. Lerman, "Design and mathematical analysis of agent-based systems," in *Formal Approaches to Agent-Based Systems*, pp. 222–234, Springer, 2001.

[21] A. Martinoli, K. Easton, and W. Agassounon, "Modeling swarm robotic systems: A case study in collaborative distributed manipulation," *The International Journal of Robotics Research*, vol. 23, no. 4-5, p. 415436, 2004.

[22] S. Berman, Á. Halász, M. A. Hsieh, and V. Kumar, "Optimized stochastic policies for task allocation in swarms of robots," *IEEE Transactions on Roboics*, vol. 25, no. 4, pp. 927–937, 2009.

[23] N. Correll, S. Rutishauser, and A. Martinoli, "Comparing coordination schemes for miniature robotic swarms: A case study in boundary coverage of regular structures," in *Experimental Robotics*, pp. 471–480, Springer, 2008.

[24] T. W. Mather, M. A. Hsieh, and E. Frazzoli, "Towards dynamic team formation for robot ensembles," in *Robotics and Automation (ICRA), 2010 IEEE International Conference on*, pp. 4970–4975, IEEE, 2010.

[25] E. Bonabeau, G. Theraulaz, and J.-L. Deneubourg, "Quantitative study of the fixed threshold model for the regulation of division of labour in insect societies," *Proceedings of the Royal Society of London. Series B: Biological Sciences*, vol. 263, no. 1376, pp. 1565–1569, 1996.

[26] E. Bonabeau, A. Sobkowski, G. Theraulaz, and J.-L. Deneubourg, "Adaptive task allocation inspired by a model of division of labor in social insects.," in *BCEC*, pp. 36–45, 1997.

[27] G. E. Robinson, "Modulation of alarm pheromone perception in the honey bee: evidence for division of labor based on hormonall regulated response thresholds," *Journal of Comparative Physiology A*, vol. 160, no. 5, pp. 613–619, 1987.

[28] G. E. Robinson, "Regulation of division of labor in insect societies," *Annual review of entomology*, vol. 37, no. 1, pp. 637–665, 1992.

[29] R. E. Page Jr and S. D. Mitchell, "Self organization and adaptation in insect societies," in *PSA: Proceedings of the biennial meeting of the philosophy of science association*, pp. 289–298, JSTOR, 1990.

[30] G. Theraulaz, E. Bonabeau, and J. Denuebourg, "Response threshold reinforcements and division of labour in insect societies," *Proceedings of the Royal Society of London. Series B: Biological Sciences*, vol. 265, no. 1393, pp. 327–332, 1998.

[31] J. C. Jones, M. R. Myerscough, S. Graham, and B. P. Oldroyd, "Honey bee nest thermoregulation: diversity promotes stability," *Science*, vol. 305, no. 5682, pp. 402–404, 2004.

[32] S. Nouyan, "Agent-based approach to dynamic task allocation," in *Ant Algorithms*, pp. 28–39, Springer, 2002.

[33] A. Martinoli, E. Franzi, and O. Matthey, "Towards a reliable set-up for bio-inspired collective experiments with real robots," in *Experimental Robotics V*, pp. 595–608, Springer, 1998.

[34] M. J. Krieger and J.-B. Billeter, "The call of duty: Self-organised task allocation in a population of up to twelve mobile robots," *Robotics and Autonomous Systems*, vol. 30, no. 1, pp. 65–84, 2000.

[35] N. Kalra and A. Martinoli, "Comparative study of market-based and threshold-based task allocation," in *Distributed Autonomous Robotic Systems 7*, p. 91101, Springer, 2006.

[36] S. Russel and P. Norwig, *Artificial Intelligence: A Modern Approach*. No. Chapter 2, Prentice-Hall Inc, 1995.

[37] P.-P. Grassé, "La reconstruction du nid et les coordinations inter-individuelles chez bellicostitermes natalensis et cubitermes . sp. la theorie de la stigmergie: essai dinterpretation du comportement des termites constructeurs.," *Insectes Soc*, vol. 61, pp. 41–81, 1959.

[38] T. Balch, "Communication, diversity and learning: Cornerstones of swarm behavior," in *Swarm Robotics* (E. Şahin and W. M. Spears, eds.), no. 3342 in Lecture Notes in Computer Science, pp. 21–30, Springer Berlin Heidelberg, Jan. 2005.

[39] G. Beni, "Order by disordered action in swarms," in *Swarm Robotics* (E. Şahin and W. M. Spears, eds.), no. 3342 in Lecture Notes in Computer Science, pp. 153–171, Springer Berlin Heidelberg, Jan. 2005.

[40] K. Lerman and A. Galstyan, "A general methodology for mathematical analysis of multi-agent systems," *ISI-TR-529, USC Information Sciences Institute, Marina del Rey, CA*, 2001.

[41] D. T. Gillespie, "A general method for numerically simulating the stochastic time evolution of coupled chemical reactions," *Journal of computational physics*, vol. 22, pp. 403–434, 1976.

[42] D. T. Gillespie, "Exact stochastic simulation of coupled chemical reactions," *The journal of physical chemistry*, vol. 81, no. 25, pp. 2340–2361, 1977.

[43] M. D. Breed, G. E. Robinson, and R. E. Page Jr, "Division of labor during honey bee colony defense," *Behavioral Ecology and Sociobiology*, vol. 27, no. 6, pp. 395–401, 1990.

[44] O. Shehory and S. Kraus, "Methods for task allocation via agent coalition formation," *Artificial Intelligence*, vol. 101, no. 1, pp. 165–200, 1998.

[45] H. Carlsson and E. Van Damme, "Global games and equilibrium selection," *Econometrica: Journal of the Econometric Society*, pp. 989–1018, 1993.

[46] W. Yoshida, B. Seymour, K. J. Friston, and R. J. Dolan, "Neural mechanisms of belief inference during cooperative games," *Journal of Neuroscience*, vol. 30, pp. 10744–10751, Aug. 2010.

[47] S. Suzuki, R. Adachi, S. Dunne, P. Bossaerts, and J. OâDoherty, "Neural mechanisms underlying human consensus decision-making," *Neuron*, Apr. 2015.

[48] D. M. Gordon, "The organization of work in social insect colonies," *Nature*, vol. 380, no. 6570, pp. 121–124, 1996.

[49] E. Bonabeau, G. Theraulaz, and J.-L. Deneubourg, "Fixed response thresholds and the regulation of division of labor in insect societies," *Bulletin of Mathematical Biology*, vol. 60, no. 4, pp. 753–807, 1998.

[50] R. M. Raafat, N. Chater, and C. Frith, "Herding in humans," *Trends in cognitive sciences*, vol. 13, no. 10, pp. 420–428, 2009.

[51] S. E. Morris and H. S. Shin, *Global games: theory and applications*. Cowles Foundation for Research in Economics, 2000.

[52] A. Martinoli, *Swarm intelligence in autonomous collective robotics: From tools to the analysis and synthesis of distributed control strategies.* PhD thesis, 1999.

[53] C. R. Kube and E. Bonabeau, "Cooperative transport by ants and robots," *Robotics and autonomous systems*, vol. 30, no. 1, pp. 85–101, 2000.

[54] D. V. Pynadath and M. Tambe, "Multiagent teamwork: Analyzing the optimality and complexity of key theories and models," in *Proceedings of the first international joint conference on Autonomous agents and multiagent systems (AAMAS): part 2*, pp. 873–880, ACM, 2002.

[55] B. P. Gerkey and M. J. Mataric, "Multi-robot task allocation: Analyzing the complexity and optimality of key architectures," in *Robotics and Automation, 2003. Proceedings. ICRA'03. IEEE International Conference on*, vol. 3, pp. 3862–3868, IEEE, 2003.

[56] M. J. Matarić, G. S. Sukhatme, and E. H. Østergaard, "Multi-robot task allocation in uncertain environments," *Autonomous Robots*, vol. 14, pp. 255–263, 2003.

[57] T. D. Seeley, "Social foraging in honey bees: how nectar foragers assess their colony's nutritional status," *Behavioral Ecology and Sociobiology*, vol. 24, no. 3, pp. 181–199, 1989.

[58] D. Fudenberg, *The theory of learning in games*, vol. 2. MIT press, 1998.

[59] K. C. Border, "Fixed point theorems with applications to economics and game theory," *Cambridge Books*, 1990.

[60] G. Camilli, "Teachers corner: Origin of the scaling constant d= 1.7 in item response theory," *Journal of Educational and Behavioral Statistics*, vol. 19, no. 3, pp. 293–295, 1994.

[61] B. Touri and J. Shamma, "Global games with noisy sharing of information," in *Decision and Control (CDC), 2014 IEEE 53rd Annual Conference on*, pp. 4473–4478, Dec 2014.

[62] N. Farrow, J. Klingner, D. Reishus, and N. Correll, "Miniature six-channel range and bearing system: Algorithm, analysis and experimental validation," in *IEEE International Conference on Robotics and Automation*, (Hong Kong), 2014.

[63] J. Klingner, A. Kanakia, N. Farrow, D. Reishus, and N. Correll, "A stick-slip omnidirectional drive-train for low-cost swarm robotics: Mechanism, calibration, and control," in *IEEE/RSJ International Conference on Intelligent Robots and Systems (IROS)*, 2014.

[64] E. Bonabeau, M. Dorigo, and G. Theraulaz, *Swarm Intelligence, From Natural to Artificial Systems.* Oxford University Press, 1999.

[65] B. Hunter and Hunter, *Statistics for Experiments.* Wiley, 1978.

[66] W. Feller, "On the normal approximation to the binomial distribution," *The Annals of Mathematical Statistics*, vol. 16, no. 4, pp. 319–329, 1945.

[67] N. Correll and A. Martinoli, "Modeling and designing self-organized aggregation in a swarm of miniature robots," *The International Journal of Robotics Research*, vol. 30, no. 5, pp. 615–626, 2011.

[68] N. Correll, "Parameter estimation and optimal control of swarm-robotic systems: A case study in distributed task allocation," in *IEEE Int. Conf. on Robotics and Automation (ICRA)*, 2008.

www.ingramcontent.com/pod-product-compliance
Lightning Source LLC
Chambersburg PA
CBHW041711200326
41518CB00001B/153